ON COMPOST

A Year in the Life of a Suburban Garden

SCOTT RUSSELL SMITH

Published by Christmas Lake Press 2024

www.christmaslakecreative.com

ISBN 978-1-960865-06-9

The names of certain individuals have been changed to protect their privacy.

Interior layout by Daiana Marchesi

Illustrations adapted by Thomas G. Fiffer from photographs by the author

Dedication

To my mother and father

Acknowledgments

I'd like to express my gratitude for, and to, my neighbors who have contributed to the ongoing community project that is my compost pile and, ultimately, this effort to chronicle its annual tending and meaning. I sincerely appreciate your kindness, generosity, and forbearance.

My special thanks to Jacoba Lawson, who was the first to realize this book was not just about raising a heap of compost but also a son. Her thoughtful reads helped turn "My Pile" into what I hope others will see as more than a how-to for environmentally conscious backyard gardeners but a guide to being a better father, neighbor, and person.

Cole, you are my inspiration. It has been my goal all along to leave our modest home ground in better shape than I found it, not just for you but for all who are inheriting a world faced with unprecedented challenges and in need of finding increasingly urgent solutions.

Select portions of the reporting that informs these pages have previously appeared in articles written for Friends of Animals, an international advocacy organization based in Darien, CT. I thank FoA for allowing me to workshop this material in furtherance of a shared vision to address the biggest contributors to climate change

caused by human activity—deforestation, animal agriculture, and fossil fuels. I am similarly indebted to Dan Woog, creator of the *06880* website, surely one of the country's preeminent hyper-local blogs. I truly appreciate the opportunity Dan has given me to be one of the voices in the shared conversation he has so ably carried on over the years. What's more, *06880* is also where I met Thomas Fiffer, publisher of Christmas Lake Press and the editor who has helped fashion *On Compost* into its finished form. Thank you, Thomas, for giving me the chance to see this quixotic venture through.

Readers will quickly recognize my profound indebtedness to many practitioners of the art and science of both writing and composting. I can only pray that all those contributors to the commonplace that is *On Compost* will find their inclusion on these pages to be, as an esteemed colleague once put it, "fair to you, true to me."

TABLE OF CONTENTS

INTRODUCTION

The compost heap I keep has long been a fixture of my life and landscape, unassuming but hard to miss. This book started as the journal of a quirky hobby, a personal pastime kept largely to myself. Lots of gardeners keep a diary, and I wanted to write about something I loved and wished to learn more about. I knew from the get-go that musing about nematodes or anaerobic bacteria wouldn't be everyone's cup of tea. But I began to take obstinate pride in elevating composting into a noble if crunchy-granola cause. The more I dug into it, the more I came to see my compost pile as a portal, a connection to issues I care about and that are fast becoming critical for us all to resolve: The extraordinary amount of food we throw away and the energy wasted in the process. Pollution and pesticides and the precipitous decline in biodiversity. How we're causing the climate to go all crazy. In time, my compost heap morphed into a backyard science project as well as a history lesson. Who knew that Darwin was so into earthworms?

I started a blog and over the course of a year detailed the tending of my pile through the splendidly changing seasons of coastal southern New England. Though only a handful of people stumbled across it, I posted updates through the next composting season. I was continually surprised by how the effort repeated itself week to week, year over year.

As I worked at home during the pandemic, the compost heap and my writing about it received renewed attention. I cobbled together the old blog posts, adding fresh impressions and seasonal recurrences. To be sure, I'm not the first to delve into the productive subject that is composting; you'll see I quote liberally on the pages that follow. Consider this book a commonplace of compost. That's an old-fashioned term for the compendiums of useful facts, observations, and ideas people have jotted down through the ages in notebooks, scrapbooks, and on parchment, if not first etched in stone. Leonardo da Vinci had a killer commonplace. They were all the rage during the Enlightenment and a staple on through the Victorian Era.

And so my book became a kind of compost heap of its own—a place where I gathered and distilled my thoughts and findings about the care my pile receives and the blessings it returns. *On Compost* chronicles the cultivation of a suburban yard as a way to highlight the larger story about how to live more sustainably and with ecological purpose. The tale I found myself telling is also about raising a son, befriending a whole neighborhood, and productively engaging with the larger world upon which my humble heap is just a speck.

My real pile has shifted locations a few times, making way for other backyard happenings—a hand-me-down trampoline trundled over from across the street, the drop-off of a ready-made tool shed, racks of kayaks, stacks of firewood. After dispensing the season's compost one fall, I hop-scotched the log-wall sidings a few feet to one side, figuring the ground beneath the old heap would be primed for planting. That garden patch is now the most profuse on the property.

This story begins—as it should—with the annual cleanup of fall leaves and continues through the frozen days of winter, the bounty of spring, and on to the heap's maturation and dispersion by late summer. It ends by coming full circle with the next buildup of autumnal organics well in hand. There are detours made and road trips taken, not always for my pile but usually for its benefit. When your life revolves around a compost heap, sometimes it takes you to curious places.

I hope you enjoy journeying through the seasons of my pile. May you find its ground as fertile as I have. If it inspires you to start composting on your own or to pitch in on a heap with your neighbors or community garden, all the better.

Now I am terrified at the Earth, it is that calm and patient,
It grows such sweet things out of such corruptions,
It turns harmless and stainless on its axis, with such endless successions of diseas'd corpses,
It distills such exquisite winds out of such infused fetor,
It renews with such unwitting looks its prodigal, annual, sumptuous crops,
It gives such divine materials to men, and accepts such leavings from them at last.

—Walt Whitman, "This Compost"

PROLOGUE

Dive Right In

It's a crisp, early November morning. I lean over the grimy edge of the parking lot dumpster, heels up and toes down, the metal lid pressed against the top button of my baseball cap. Plucking a pendulous bag from the jumble of trash below, I raise it to arm's length, like a trophy fish. Twenty pounds easy. No drippy leaks or cast-off paper cups, lids, or stir straws poking through the thin white plastic.

It's a keeper.

I set the bag, still warm to the touch and ripe with the dank, roasted aroma of freshly spent coffee beans, on the floor well behind the driver's seat. Backing away from the rusted bin, I'm relieved that no aproned barista is running out the door asking me what the heck I'm doing.

Truth is, I'd do most anything for my compost pile. Even if that means getting—I see with a glance at my lap—a smear of grease on my good leather jacket. Oh, well.

It's the middle of fall in Westport, Connecticut, an artsy, affluent New York commuter suburb along the northern shore of Long

Island Sound. Over the past few weeks, the trees that surround my small home on its flat corner lot have largely shed their leaves. I've raked the colorful, crinkly flakes into piles and hauled them by the blanketful over to the log-walled compost heap I keep in the back corner of the yard.

At this time of year, the pile is head high and a broad-jump deep and wide. It's the copious conclusion to a burst of green growth that culminates in "leaf-peeping season" here in southern New England. To me, the autumnal leaves of hardwood trees are more than a tourist draw or scenic perk of living in a four-season clime. I see each leaf as a bank slip of carbon and other minerals and nutrients just waiting to be deposited as new gains over the coming year.

But for my pile to cash in, all those dried-up leaves need a catalyst, something to kick-start their conversion into new, living soil. Rich in nitrogen, the twenty pounds of recycled coffee grounds rescued from the trash will deliver the same jolt of energy to the heap of leaves that they gave to scores of caffeinated customers earlier this morning.

At least that's how I rationalize my scrounging. After all, coffee is one of the most-traded commodities on the planet, with 23.2 billion pounds processed each year. That's a whole lot of coffee and a whole lot of spent grounds. Lightening the coffee shop's dumpster by a hefty bag saves the local garbage hauler from having to truck away that much more organic waste to a distant, already overstuffed landfill.

I'm thinking locally, acting globally. As master composter Mary Tynes puts it, "Environmental protection doesn't just happen on

the other side of the world. Our first responsibility is to care for the patch of soil on which we live. People who understand soil and how Nature replenishes it are able to make more responsible choices on both small- and large-scale environmental policies, and wider socioeconomic issues."

Garbage in, garbage out? Not with my pile. The bits and pieces of digested life and matter that make up this backyard compost heap, in time and with some tending, always reconstitute themselves into something wonderfully new and useful. My pile is my touchstone, a wellspring of life that nourishes the garden and me as I nourish it. Some count sheep to drift off to sleep; I turn over shovelfuls of compost in my head. After all, what's a brain but an organic repository for gathered thoughts and things to be processed, to be... composted? Visualizing, X-ray-style, what's in my pile, sorting through its unseen layers, soothes my soul.

"Composting" is what author Natalie Goldberg calls this filtering process. "Our bodies are garbage heaps: we collect experience, and from the decomposition of the thrown-out egg shells, spinach leaves, coffee grinds and old steak bones of our minds come nitrogen, heat, and very fertile soil," she explains in *Writing Down the Bones*. "Out of this fertile soil bloom our poems and stories. But this does not come all at once. It takes time. Continue to turn over and over the organic details of your life until some of them fall through the garbage heap of discursive thoughts to the solid ground of black soil."

The longer and more deeply I explore it, the more firmly my pile remains *terra incognita,* a Rubik's cubic yard of shape-shifting

organic matter that defies understanding. Still, I try, if only for the exercise. It is the nexus of the physical and psychic place where I spend my most agreeable waking hours—the backyard. For this reason alone, the mysterious heap is worth getting to know.

I am comforted by the fact that there is a rich history of such landscape navel-gazing, from Henry David Thoreau, who "went to the woods because I wished to live deliberately, to front only the essential facts of life, and see if I could not learn what it had to teach," to the Irish poet and novelist Patrick Kavanagh: "To know fully even one field or one land is a lifetime's experience. In the world of poetic experience it is depth that counts, not width. A gap in a hedge, a smooth rock surfacing a narrow lane, a view of a woody meadow—these are as much as a man can fully experience."

As a construct, my pile may have modest value that extends beyond this one backyard. "People *exploit* what they have merely concluded to be of value, but they *defend* what they love, and to defend what we love we need a particularizing language, for we love what we particularly know," writes Wendell Berry. I want to particularly know my compost heap. I will plunge into it deeply, turn it over and over again, and, in time, reap and share its rewards.

Writing a century ago in *The Gardener's Year*, Karel Capek unearthed the essence of what draws me to my backyard garden and what keeps me there: It's not the showy blossoms or ripening fruits, it's something much more basic, indeed, grounded.

"While I was only a remote and distracted onlooker of the accomplished world of gardens, I considered gardeners to be beings of a peculiarly poetic and gentle mind, who cultivate perfumes of flowers listening to the birds singing. Now, when I look at the affair more closely, I find that a real gardener is not a man who cultivates flowers; he is a man who cultivates the soil. He is a creature who digs himself into the earth and leaves the sight of what is on it to us gaping good-for-nothings. He lives buried in the ground. He builds his monument in a heap of compost."
– *Karel Capek*

NOVEMBER

Raking It All In

The capacity of my pile to absorb ever more never ceases to amaze me. It is not so much a bottomless pit as a topless one.

The average mature tree has some 200,000 leaves; the annual leaf mass in an acre of mixed hardwood forest can total two tons or so. Heavy lifting? Hardly, if you pace yourself. A single leaf, untouched by rain or morning dew and dried to a crisp by the sun, weighs next to nothing. A pile of leaves amounts to more air than anything else. As if to prove the hack golfer's axiom, a tree really is 90 percent air, even—especially—when its canopy of leaves is splayed flat across the ground.

My favored form of fall cleanup is to rake up a slouchy heap of leaves, lay an old bedsheet beside it, then scrape the leaves over until just the four corners of the sheet are visible. I gather the points together, sling the makeshift bag over a shoulder, and haul it to the heap. If the leaves are wet with dew, I drag the damp sheet across the ground. For the most part, it's easy pickings. Every few days throughout fall, I make time to collect and deposit loads of this ephemeral fluff across the top of my pile. It spills over the sides, strains against the wire fence stretched across the back,

and cascades down the open front. Each leaf finds its own angle of repose. It's a rhythm I keep throughout the season.

My father introduced my brother and me to this way of yard duty. We lived in Kentucky at the time, on an acre lot with a parkland spread of mature hardwood trees. We'd rake the leaves into big piles, spread a painter's drop cloth across the just-swept ground, raking and kick-walking the leaves high onto the sheet. This process works only if you have a nearby place to dump the leaves. We did—across the street was a wooded ravine that sloped down to a creek. Gathering together two corners each, like a king-sized hammock, we'd drag the load across the street and unfurl it down the steep slope that began just past the pavement.

I've used the same old double-bed sheet in my much smaller yard for the past four or five years. It's battle-scarred, ripped by sticks, stained by mud and tannins, and sports a duct-tape patch over a tear in the middle. It is still serviceable, and using it is much more appealing than the prospect of trying to stuff a dozen or more tall brown paper bags for each cleanup. For one, it's hard to grasp a mess of leaves by hand. And jamming leaves down the throat of those bags is an exercise in frustration, even if you don't end up ripping one with a stray branch or wayward rake tine.

In volume, the size of my pile is limited by two rows of upright logs on either side. They are about eight feet apart, with a length of wire garden fence stretched between the two tallest logs at the back. These log pillars, seven on one side, eight on the other, are set in ascending order of height. It's a rough staircase that's useful when dragging a load of leaves up the open, sloping front to deposit on

top. In years past, when my son, Cole, was small enough, he and his friends would clamber up the twin rows of logs and delight in jumping in and getting swallowed whole. Their antics would flatten and smoosh the air out of my pile, allowing me to add even more. My boy, now in his final year of high school, has long outgrown the joy of jumping into leaf piles, and as I've stepped up my scavenging efforts, the heap is no longer fit for little kids to free-fall through. These days the coffee grounds, kitchen scraps, and seaweed make it my playground alone.

If the weather and my office schedule cooperate, I put in an hour or two after work several times a week—at least until daylight saving time robs me of that extra hour of sunlight. It's enough time to rake up a load of leaves from under a tree or from along the street gutters that bound my corner lot. Weekends are for fuller cleanups. Some people hate the very idea of raking leaves, or yard cleanup of any kind, and outsource the whole process. They're happy to have the town take bags of leaves off their hands or to hire a lawn service and be left with a blown-clean yard. I see either option as a colossal waste of resources, both municipal and my own. More than 10 million tons of leaves and other yard trimmings ended up in landfills in 2018, according to the Environmental Protection Agency; that's over 7 percent of all waste thrown away. To take all this material away from its source, midstream in the lifecycle, seems to me a clear-cut loss. No wonder companies sell so much fertilizer each spring, after so many consumers spend so much to have it taken away from their homes each fall.

I figure about half of each season's leaves go straight into the heap whole, raked from the lawn and street. The rest get left in the garden

beds or mulched by the lawn mower to be added to the heap or left to disintegrate back into the grass when the hopper is full. I admit to envying, to a degree, the lawn maintenance crews that use a vacuum hose to scarf up leaves and shred them to pieces into a big wooden box in the back of a two-ton truck. What use I could make of all that finely chopped leaf litter! But my home-style leaf cleanup works for me. I find raking a pleasant excuse to spend time outdoors doing meditative busywork, crossing a chore of home ownership off the list without having to write a check.

For a gardener, raking provides a tactile connection to the ground. I touch virtually every square foot of my property at least once a year. Sweeping the lawn clean of fallen leaves is a peculiar cross between vacuuming the carpet and giving the dog a scratchy rub. The rhythmic, repetitive movement allows the mind to wander. I'd rather listen to the tines of the rake scraping across the ground than the dusty, noxious blast of a leaf blower. There's something off-putting about using a noisy, highly polluting two-stroke combustion engine to recycle green energy from my yard. I also like the exercise raking affords, the basic movement a combination of a hockey player taking a slap shot and a conductor waving a baton. There is an art to raking, after all.

Spades take up leaves
No better than spoons,
And bags full of leaves
Are light as balloons.

I make a great noise
Of rustling all day

Like rabbit and deer
Running away.

But the mountains I raise
Elude my embrace,
Flowing over my arms
And into my face.

— Robert Frost, from "Gathering Leaves"

Poets don't write odes to leaf blowers. What's more, the high-revving motor of my old Toro mower and two-stroke blower spew a worrisome amount of noxious fumes. According to the California Air Resources Board, operating a gas leaf blower or mower for an hour can create as much smog-forming pollution as driving a Toyota Camry 1,100 miles. For sure, my next mower will be electric. Two of my neighbors have already switched over to battery-powered mowers and blowers, a particularly easy decision given the relatively small yards in our neighborhood. That's good news for our lungs and music to any suburbanite's ears.

A lawn mower of whatever stripe does a good job mulching volumes of leaves into fractions of their former selves. Chopped-up leaves are manna to my pile. The whirring blades slice and dice each leaf into bite-sized meals for all the things that want a piece of it. Decay usually begins at the frayed edges. Composting is firstly about deconstruction: of a whole leaf, banana peel, or eggshell into ever-smaller increments, mostly by something eating something and pooping it. Then you die and become a meal for something else, usually smaller. Off the top of my head, I couldn't give you the technical definition of entropy, but I have a feeling that my pile is a pretty good example. Things fall apart in my pile according to laws governing biology and thermodynamics—along with some magical dark arts, it often seems.

Like Hermione Granger's magic bag, a compost heap that appears stuffed is always able to accept a new gathering of leaves—especially when they're mulched. A grass-catcher bag or two of the dry, dusty stuff makes a fine topping when I'm cleaning up the yard ahead of a coming storm. Mulched leaves don't scatter in the wind; they soak up any rain that may come and, spread across the top, give the heap a tidy, manicured look.

The best, most productive way to condense a pile is to just add water, and all it takes to shrink my pile is to wait for a good drenching rain. Water is weight, and in the right proportion, H_2O is also the catalyst for all the energy my pile will consume and create over the next few months. "Rainwater is the best kind to put on compost. It picks up lots of oxygen, minerals and microorganisms as it falls through the air, giving your compost an added boost," writes Stu Campbell in *Let It Rot! The Gardener's Guide to Composting*.

"The amount of water in your compost pile is fairly critical, but you have plenty of leeway in which to work. If the moisture content is much greater than 60 percent, you run the risk of having an anaerobic pile; if it is much less than 40 percent, organic matter will not decompose rapidly enough because the bacteria are deprived of the moisture they need to carry on their metabolism. Of course, you can't monitor the percentages, so in general try to make sure the materials in your pile have the moisture content of a well-wrung sponge." – *Stu Campbell*

Throughout fall I've taken care to add as much moist material as possible—grass clippings, wet leaves, musty seaweed. But I know from the vapors beginning to rise from the top of the heap on frosty mornings that the steam engine that is my pile is boiling off much of the water it has stored up inside. If the weather is dry—as it's been this fall—I reach for the garden hose. You can see the weight of the water sucking the pile back into itself. If I'd just ingested a half-dozen trees' worth of dry, waxy cellulose, I'd be thirsty too. I leave the hose on for five or so minutes, poking here and there. Figuring a gallon of water weighs eight pounds, I'm adding nearly a hundred pounds, say a bathtub's worth. It's always a mystery where the water goes and what unseen path it takes through the matrix of leaves. Usually, by the time I come up with a guess, a trickle of water is dribbling out to soak my shoes.

Gourmet Beginnings

A gardener is very much an editor—from planting the seed of an idea to pruning prose that has run amok. Both writing and gardening take shape in the mind and develop through careful and consistent attention. That sensibility has played out over my career, largely spent writing and editing for a mix of magazines, mostly of the how-to variety. My vocation as a writer and editor and my avocation as a gardener and composter go hand in hand. I often mull over writing projects while busying myself with the pruning, curating, and transplanting that keep a suburban gardener preoccupied. "I have never had so many good ideas day after day as when I worked in the garden," claims author and composer John Erskine. I also sometimes plot out backyard projects while idling myself at work.

Making compost is a creative, unscripted act, yet its production plays out in a time-honored form and fashion. My pile begins as a load of raw if purposeful rubbish, a rough draft that with time and effort is refined into a finished product put to immediate use. A heap of fresh compost is very much a magazine; at its root, the word refers to a collection or storage location, like for gunpowder at a military depot. Consider my compost pile the *Guns & Ammo* of gardening.

I've kept a backyard compost heap since my days as a staff editor for a food magazine in Los Angeles in the mid-1990s. It was a wonderful job in many respects, chief among them the twice-daily tastings in

the test kitchen. Every recipe that ran in the magazine, and then some, was first prepared in-house by our chefs, with assists from other staffers and guest editors. The tastings were held three days a week, the first at 10 a.m. and the second at 2 p.m. I'd hang out in the test kitchen as much as I could, watching and listening and smelling all that went into their work while avoiding the stack of recipe transcripts and manuscripts in my inbox. This was in the early days of modern comfort food. The artery-clogging, cholesterol-laden recipes of old were being replaced by healthy Mediterranean menus, featuring lots of vegetables simply prepared and seasoned. Gone were the rich sauces; fresh herbs and garlic appeared in their place. Lots and lots of garlic. I ate amazingly well from 9 to 5 and couldn't get a date at night for the better part of five years.

I don't recall how the subject of my wanting the trimmings the cooks threw out each day came up. But once they knew of the modest compost heap I'd started in the side yard of the duplex I rented at the base of the Hollywood hills, my colleagues happily loaded me up with all the kitchen scraps I could take home at the end of each testing day. It was gourmet stuff—floppy green carrot tops and big bottoms of fennel bulbs. Pounds of flicked potato peels, whole volumes of papery onion skins. Lots of shrimp shells, as I recall. All in all, enough, usually, to fill two grocery bags every test-kitchen day. Each issue of the magazine included about 100 recipes, which, every month, found their way into a million or more kitchens. That's a lot of food scraps.

My early contributions to composting were barely a drop in the bucket. According to Jonathan Bloom at *Wasted Food*, a staggering 95 percent of the food waste produced in the United States that could be composted actually ends up going into a landfill or incinerator. In fact, food is the single largest component taking up space inside U.S. landfills, accounting for 22 percent of all municipal solid waste created each year. "That equates to 325 pounds of waste per person—the equivalent of 130 billion meals," according to the *Food Waste in America in 2023* report from Recycle Track Systems. But even the minuscule amount of food and other organic material (including paper products and yard trimmings) that is composted or recycled makes a huge difference. Taking one year as an example, diverting these materials from landfills "prevented the release of approximately 186 million metric tons of carbon dioxide equivalent into the air in 2013—equivalent to taking over 39 million cars off the road for a year," according to an EPA report Bloom cites. It didn't occur to me at the time, but I suppose my LA compost pile offset a fair share of the smog I created driving back and forth to work each day, helping to produce a monthly magazine made of many acres of wood pulp.

What happens after we toss those day-old leftovers into the trash can? Most of us don't think about it. But as David Owen points out in *The Conundrum*, rotting food is the main source of the methane—an especially worrisome greenhouse gas—that leaks from landfills. That's not all:

"When we throw away food we don't just throw away nutrients. We also throw away the energy we used in keeping it cold as we lost interest in it, as well as the energy that went into growing, harvesting, processing, transporting, and preparing it (assuming we got that far), along with its proportional share of our staggering national consumption of fertilizer, pesticides, irrigation water, packaging, and landfill capacity." – David Owen

My Spanish-style duplex sat perched on a knoll just below the Griffith Observatory. On the edge of the back patio a tall Ponderosa pine, its lower branches trimmed, stretched its upper limbs outward to form a wide, sculpted canopy. It was a gorgeous tree, especially in the evening when the sun set over the Pacific. The colors would wash across the Los Angeles basin and light the tree from underneath, making its orange-red bark glow and waxy green needles sparkle. A patch of ivy covered the slope below it. And there, in the bottom corner of the yard, I carved out my first compost pile, digging steps into the terraced hillside to reach it.

The rest of the backyard was taken up by two old olive trees that shed long, slender leaves and a rich rain of black olives that slowly air-dried to dusty pits. In their shadow grew a rosemary bush so big that when I trimmed it, I saved the branches to use as skewers for kabobs. Alongside the house, a sliver of grass bordered the walkway to the street. This tiny lawn gave way to a small rose garden perched above a huge hedge of brilliant magenta bougainvillea that fronted the sidewalk. Though small and set on a hillside, the yard produced enough throughout the long California growing season to keep my compost heap in business. I especially liked scooping up piles of the Day-Glo bougainvillea petals, as thin as the breath strips you put on your tongue and just as fast to melt away.

Spoiled by the chefs in the test kitchen, at times my pile was more kitchen scraps than yard refuse. It was a turbocharged stew of vegetable matter, with just a few rakefuls of pine needles, prickly live-oak leaves, grass clippings, and homegrown rotting olives mixed in. My landlord was pleased that I took ownership of the yard. My downstairs neighbor, Alix, a British émigré who worked as a paralegal, was also happy to have me puttering about, producing a steady supply of fresh-cut flowers for her apartment.

Being a monthly magazine, we worked off an editorial calendar set several months ahead of real time. In September we tested our Thanksgiving menus, which meant that our fattest issues hit at the tail end of California's dry season—just before winter rains began. The test kitchen worked overtime in those months, producing turkey after turkey to taste, providing me with all the trimmings to take home. Pound for pound, there wasn't a more fortuitous compost pile in all of Los Angeles. My gourmet compost pile helped turn a rented patch of compacted, hardpan dirt into a lush backyard oasis. When the fall rains came, the garden soaked up every drop, and the roses and bougainvillea and rosemary thrived in the California sunshine.

And I became a composter.

Composting is a pastime, a passion, a pursuit. Tending to a backyard compost heap calls for care and feeding to keep it active and in good health. It's like having a pet, and I'm devoted to it. Like any pet owner, I am sensitive to comments about it. One being, "Ew, a compost pile? With all your garbage in it? Won't that just attract rodents, like rats?" Well, yes—but it's not my pile's fault; it's just doing what compost does. It's its nature. It *is* nature.

We share our suburban landscapes with all kinds of critters, welcome or not. My Los Feliz neighborhood was near rugged Griffith Park. All kinds of urban animals would wander over, chiefly because Alix set out bowls of kibble and water on the front stoop for her big black tomcat. Coming home late from work one evening, I bent over to pet what I thought was her cat crouched beside the dishes, only to realize the backside belonged to a skunk. My hillside compost pile was also a wildlife draw. One cool morning I dug into the heap and surprised a possum that had carved a nesting spot in its warm flanks. The cat gave these wild visitors a wide berth, especially the raccoons. All feared the coyotes that roamed the Hollywood hills at night.

Then Wilbur arrived. A Christmas present for Alix from a benighted suitor, Wilbur was an adorable little Vietnamese pot-bellied piglet with a red bow tied around his bristly neck. Micro pigs were the trendy pet back then, supposedly as smart as a dog, easily house-trained, and with a sense of humor—almost. Plus, this breed was promised to stay cute and cuddly. The pig pranced around Alix's ground-floor apartment like a toy poodle on high heels, hooves clicking along the wood floors. Wilbur used his snout like the mini-trunk of an elephant, always nosing around, sticking his frizzy pink nozzle into your side for stroking, sniffing out anything and everything. And that was the problem. The piglet had one button—the on switch—and he was indefatigable in his search for one thing and one thing only: food. He was a snout attached to a stomach set on cloven feet.

On occasion, I'd come home from work before Alix, and I would hear the little piggy downstairs, rooting around. He'd learned how to nose open Alix's kitchen cabinets and delighted in licking and

kicking her pots and pans across the apartment. If a molecule of a food particle, even a memory of a meal, remained on the pot, the pig would lick it until the Teflon wore off. Alix put childproof locks on all the kitchen cabinets and secured the cat-food bowls. She took the piglet for walks on a leash around the neighborhood. In time, she let Wilbur walk about the yard, as it was mostly fenced and fairly private from the street. He quickly sniffed out my compost pile. At first, I didn't really mind. Wasn't it entirely natural that a pig and a compost heap were made for each other? I was prepared to share my bounty, as there'd always been enough to go around. Besides, the pig rooting through my pile did help turn it over a bit. But soon, Wilbur's relentless pursuit of all things edible overwhelmed my little heap. It wasn't a fair fight. Any bag of fruit and vegetable scraps I added to the mix was gone by the next day. My pile became his pigpen, his feeding trough.

And he got bigger and bigger, rather than staying cute and cuddly. Wilbur put on pounds each week, and before long had grown to over 100. His personality, over-hyped to begin with, grew more single-minded. All he wanted was food. I grew resentful and stopped bringing home leftovers from the office. I allowed my pile to devolve into, well, a pigsty, more dust than loamy dirt in the making.

Then one day I got home to find Alix in a proper tizzy. She'd left the pig outside rooting around the yard while she was inside her apartment, then found him gone. She printed up "Missing Pig" posters and stapled them onto telephone poles around the neighborhood. The ABC-TV affiliate station was just a couple blocks away, and a news producer saw the flyers and sent a TV crew over

to interview her. "Pignapped! Live at 5!" led the evening newscast. Alix, the beautiful, big-eyed British ex-pat, was tearfully persuasive in pleading to be reunited with her pet pig. Sure enough, some hours later, the station got a call from a viewer in East Hollywood, who had seen the pig in a nearby backyard, the unwitting guest for a weekend barbecue in the works. Wilbur was safely returned home.

I moved a while after, taking a new job on the other side of the country. Alix and I stayed in touch, and a few months later she called to say that her cute little Christmas gift had grown to 150 pounds of pure surliness. She was sending Wilbur to an animal rescue shelter on the outskirts of the San Fernando Valley, where she hoped he would have a long, happy life. I didn't miss that pig, but I sure did miss that compost pile.

New Home Base

I relocated from Los Angeles to Connecticut, landing a better gig with a bigger magazine, this one devoted to golf. There would be no food scraps to take home, but I was looking forward to a change of scenery. I plunged into marriage, then homeownership, buying a tidy little cape in Westport, with a sizable yard of trees and grass, not far from Long Island Sound. At last, I had a compost pile I could call my own.

We had a son, and married life was nice—while it lasted. Then things fell apart, and I ended up buying a smaller, cottage-style house nearby for my five-year-old boy and me. It was an old widow's home, with a seriously neglected yard. The third-acre lot was studded with tall trees, including several types of maple, a large sycamore, a pair of overgrown mulberry trees, a big white pine, a pretty tulip magnolia in front, and an ancient, bedraggled willow tree in back. I knew I would buy it the moment I pulled into the driveway to meet the realtor. I was sold on the yard, a reclamation project that I knew would keep me preoccupied while I rebuilt my own life. Later, I heard from a neighbor that the woman of the house once enjoyed gardening, but after losing her husband and contracting Lyme disease, she gave up on maintaining the property. As she aged in place, a shut-in, the invasive vines and trash trees slowly took over, encroaching from the tree-lined edges of the yard, rolling over her garden borders

until a narrow moat of grass was all that separated her house from a suburban jungle. When I moved in, the property was the neighborhood eyesore. I couldn't wait to resurrect the yard from decades of neglect and make it my own.

Closing on the house in May, I spent that summer clearing the property of twenty years or more of unchecked growth, hauling away truckloads of brush and vines. Most of the plantings in the yard, I came to realize, hailed from distant lands. "Ornamentals," the nurseries call them, as if plants are *objets d'art* for display and not integral to a functioning landscape. I suppose these exotics were considered status symbols when the property was first landscaped in the 1950s. Without natural predators or pathogens to keep them in check and lavished with attention—not to mention massive amounts of Miracle-Gro (introduced in the late 1940s)—these exotics prospered, and no doubt gave many a gardener the idea they possessed a preternatural green thumb. The biggest immigrant winner was the lowliest plant of all, as grass lawns became the nation's largest irrigated crop by area. (Trick question: Where does Kentucky bluegrass hail from? Try North Asia and Europe.)

"Rich people always found rarity and ephemera completely intoxicating, and the tendency for the other 99 percent to chase after the novelties of the wealthy has had lasting effects on the way we thirst for plants today," writes botanist Heather Arndt Anderson for *Sunset* magazine.

One of the first projects I tackled was removing a massive burning bush that dominated the front corner of the yard, blocking sightlines of cars rounding the bend. With a boy just learning to

ride his bike without training wheels, hacking it all the way back was an easy decision, if no small task. Besides, its removal gave the tulip magnolia more room to breathe. Though the flaming red of its fall foliage is eye-catching, the import from East Asia, from the Euonymus family, grows prolifically, spreads wildly, and does next to nothing for native wildlife. Same with the infernal rose of Sharon, a showy late-bloomer that had self-seeded with abandon. Originally from China and long the national flower of South Korea, *Hibiscus syriacus* came to these shores from gardens in the Middle East. The orientalism of my yard extended to rhododendron, azalea, hosta, forsythia, privet, japonica, wisteria, and mulberry, even the pachysandra that covered the ground around the house. I practically needed a passport to garden.

Spending so much time outdoors helped me get to know my neighbors, who would stop by to praise—and appraise—my efforts at overhauling the blighted mess. Grubbing out the tangled brambles of Japanese barberry, a particularly nasty invasive, from the back corner of the yard, less than thirty feet away from my neighbors' houses on either side, I came across tramplings and scat from deer that had overnighted there in seclusion. I also had to encourage the fat and happy groundhog who lived under the back porch to take up residence elsewhere.

After a summer's worth of sweat equity, the bones of the property were revealed, and they were good. A corner lot, yet not quite square, the yard has what English garden creator Vita Sackville-West called "minor crookedness." Plotted from an onion farm that was developed in the postwar years into a modest neighborhood of capes and split-level ranches, the yard slopes from the road in front about a foot in grade. The back corner is the lowest point, tending toward the mucky. The neighborhood is less than a mile from Long Island Sound and just a few feet above the mean high tide line, which means that in wet times the water table rises nearly to ground level.

Chris, a friend in the tree business, tackled the trees that needed to come down. There was the pair of old mulberry trees that draped over two sides of the house and carpet-bombed the roof with purple berries; an invasive Norway maple fatally warped by hairy tentacles of poison ivy; and a bigger, rotted swamp maple standing at the center of the new grass lawn on which I envisioned playing catch with my young son. Dominating the backyard was an old weeping willow, a good three feet thick at the base. It had three main branches, each lopped off about twenty-five feet from the ground. Years of second growth had sprouted from the topped ends, giving the tree a sawed-off, if still majestic, crown. It was a dramatic sight, thickly cloaked in heavy strands of English ivy. As I worked my way across the rest of the yard, rooting out truckfuls of brush, I considered keeping the relic, daydreaming of an elaborate tree house constructed atop its thick trunk and tripod arms.

"The most noteworthy thing about gardeners is that they are always optimistic, always enterprising, and never satisfied. They always look forward to doing something better than they have ever done before." – Vita Sackville-West

Native to central Asia, willow trees are considered second-class citizens of the modern suburban landscape—fast-growing but unruly, messy, and weak, with invasive roots to boot. They generally don't age well. I suppose the old widow had the money to trim it but not the cash or will to take it down entirely. Likewise, at first I resisted Chris's entreaties to put it out of its misery. Cutting down the whomping old willow would nearly double my tree-clearing bill. I got a deal on the maple because it made good firewood to be hauled away as logs, but the soft, spongy wood of the willow wasn't good for anything. It would cost a small fortune to haul off, even if you could figure out a way to load it into a truck. But Chris worked out a deal, and I came home from work one day to find it in massive, chopped-up pieces.

Luckily, I was just about ready to construct a new compost heap, and those chunks of willow created a perfect home base. I rolled several sections to the corner of the yard, upending two of the biggest pieces about eight feet apart. I then hoisted two more sections on top of each of them, so that two logs stood as chest-high pillars. I stacked two twinned smaller logs next to the first pillars, pleased to find them about six inches lower than the anchor logs. For the third row I paired up two other sections, also about six inches shorter than the stacked logs before. I finished with two squat logs of park-your-butt height, creating a wooden crib with twin barked sides that stepped from two feet high to about four feet. I nailed a section of wire garden fence across the eight-foot gap between the two rear columns to complete a decent, if rustic, three-sided enclosure. Then I filled the empty space with its first batch of leaves, dirt, and debris left over from the cleanup.

Now I had a new patch of ground with plenty of green grass for my son to play on, freshly prepared garden beds to plant the coming spring, and a sturdy new home for my compost pile. By the time the leaves of the trees left standing began to rain down upon my newly seeded lawn, that first flush of yard waste was well on its way to being cooked. Seasons later, the willow logs are now encrusted with fungus and molds and sprout mushrooms after rains. They look like old pilings, rotting away as they age in place. I've had to replace several with newer logs, but they've done their job containing my pile, and adding to it. The wooden walls harbor billions of fungus spores and bacteria that launch themselves into each year's new pile—just like my son once did.

Falling into Place

It's a Saturday in the lingering warm autumn days of mid-November, the weekend before Thanksgiving. With a drenching rain due to arrive on Sunday, it's high time to do the final fall yard cleanup and fully stuff my pile.

Leaf season in these parts lasts about six weeks, beginning in early October when the first leaves flutter down from the tops of the tall hardwood trees that ring my yard. Some fall on the lawn and garden and gutters; others are blown to the curb by passing cars and delivery trucks. I focus on the street leaves first, as they're easy to rake up in piles to transport to the heap. It keeps the curbside gutters and storm drains from clogging. Most autumns, the lawn kicks in with a burst of renewed growth, spurred on by warm days, cooler nights, and fall rains. The kitchen garden contributes its diminishing returns until the first hard frost, usually in late October. The perennials, small shrubs, and ornamental trees that crowd my garden beds hang on to their color and berries the longest, nowadays past Thanksgiving.

All this seasonal growth is dispatched according to its schedule, and my own, which turns out to be exactly what we both need. The process of putting my lawn and garden to bed for winter and filling up my pile makes this one of my favorite times of year. It's certainly the most active, a race to harvest the year's outpourings and put

them to a productive end use. And what better time to be outdoors and soaking in the daylight of an Indian summer day before the long New England winter sets in?

By bulk, green grass and brown leaves are the two main components. It's easy to add heaping layers of each—an after-work session of leaf raking, followed by a weekend mow. The most productive compost pile is one with a healthy proportion of fresh green things and dead brown things, in about a one to three ratio. A pile of cut grass cooks itself into a rotten mess. A pile of leaves just sits there. Mixed together, they have real chemistry. At this early stage of the pile's lifecycle, it is always more leaves than anything else. The more freshly rotting greens that I add, the hotter it will cook through the winter months. And this mass of leaves and compostable whatnot will boil down into a finished batch of loamy new humus sooner as well. Here's how it plays out in my backyard each fall:

If I rake up a batch of leaves one day after work, the next day I'll take the dog for a run on the beach. I bring with me a plastic barrel—the kind you fill with ice and set a beer keg in—and scrape up a bucketful of seaweed from the high-tide line to spread over the leaves. Later in the week, I'll hit another spot in the yard for more leaves, then add buckets of kitchen scraps, from both my house and my neighbors'. Sometimes this will include a shopping bag full of shavings and pellets from their rabbit hutch. Or I'll stop by Starbucks for a supply of spent coffee grounds. There aren't many rules to building my pile, more like guidelines—and opportunities. I see value in every garbage can and recycling bin. Earlier this week, the older couple next door asked if I could take the Halloween pumpkins from their front stoop. They'd served their decorative

purpose. Rather than waste their composting potential, I chucked 'em in with the rotting husks of the jack-o'-lanterns Cole and I had carved.

A member of the *Cucurbitaceae* family, which also includes squash, cantaloupes, cucumbers, watermelons, and gourds, the pumpkin has been cultivated for a thousand years or more, first by Native Americans in the Southwestern United States. Today many are grown near big cities to cater to the Halloween market. According to the researchers at the Penn State Extension, one billion pounds of pumpkins are harvested in the U.S. each year. Though I'm heartened by the fact that the other major holiday decoration—the Christmas tree—is often recycled, the fate of millions and millions of these pumpkins is less certain. It's why I like adding their orangeness to my pile late each fall. Pumpkins are, like most vegetables, more than 90 percent water. Despite their heft, or perhaps because of it, their remains disappear without a trace in my pile, but add measurably to it.

I'm sure other composters in other places have their own localized routines and recipes. I've seen lists of all the things you can compost, and it's an impressive array, from dryer lint to hair swept from the floor of the barbershop. In *Let It Rot*, Stu Campbell includes such items as feathers (high in nitrogen and phosphoric acid) and tobacco dust and stems (a rich source of potash). I have no interest in adding cat litter or dog doo to the heap (nor should you) but am pleased to know that Zoo Poo, "made from elephant, rhinoceros, and other herbivorous animals' poo," is on Nicky Scott's list in *How to Make and Use Compost: The Ultimate Guide*.

Karel Capek's approach in *The Gardener's Year* mirrors my experience. "There are times when the gardener wishes to cultivate, turn over, and compound all the noble soils, ingredients, and dungs. Alas! there would be no space left in his garden for flowers. At least, then, he improves the soil as well as he can; he hunts about at home for eggshells, burns bones after lunch, collects his nail-cuttings, sweeps soot from the chimney, takes sand from the sink, scrapes up in the street beautiful horse-droppings, and all these he carefully digs into the soil; for all these are lightening, warm, and nutritious substances."

I once visited a work colleague who lived in a beautiful apartment in a stable house at a Connecticut estate and, much to her bemusement, left with a bucket of horse manure. The horses she lived above were worth millions, and their droppings were like gold to me. Recently I composted most of a bag of cat food—though unappetizing to Tuffy, our resident mouser, the chow will be catnip to the worms and microbes in my pile. The silliest thing I will admit to adding is a set of fingernail clippings (high in protein-rich keratin). My pile makes work for idle, if manicured hands.

The fall season starts with the autumnal equinox in late September, and back when Henry David Thoreau was writing near Boston, this was when the color change was beginning to peak.

"We only have to read Thoreau to know that climate change is pushing the changing colors later into the year. In his 1862 essay, 'Autumnal Tints,' the naturalist wrote: 'By the twenty-five of September, the Red Maples generally are beginning to be ripe. Some large ones have been conspicuously changing for a week, and

some single trees are now very brilliant,'" reports Craig S. Smith for *The New York Times*.

Now, the timing of the leaf drop is increasingly up in the air. Smith quotes Richard Primack, a biology professor at Boston University: "'In general, peak leaf color in Concord and the surrounding Boston area for these maples is now more typically a week or two later' than what Thoreau observed."

Looking at autumn color in a central Massachusetts forest studied from 1993 to 2010, Smith predicts that "the duration of the fall display would increase about one day for every 10 years." That's a week per lifetime. I am cheered to learn that by composting I am doing my part to slow the lengthening season, to prevent winter from becoming the new fall. "If we protect and sustainably manage soils," Ronald Vargas of the United Nations Food and Agriculture Organization tells biocycle.net, "we can combat climate change."

My compost heap is a very modest offset to the magnitude of soil that is lost each year across the living skin of the earth. Topsoil in North America erodes by an average of four tons per acre annually, and a whopping 36 billion tons a year washes away worldwide. That's alarming—and not just to food producers. Soil can trap huge quantities of carbon dioxide in the form of organic carbon and prevent it from escaping into the atmosphere. Unfortunately, soil is now eroding up to twenty times faster than it is being developed. In the famously fertile Corn Belt of the U.S. Midwest, 30 million acres—one-third of the entire area—has completely lost its carbon-rich topsoil.

In *The Soil Will Save Us*, Kristin Ohlson writes about reclaiming her worn-out suburban lawn through the use of leaf mold and compost, then uses the experience to tell the larger story of "How Scientists, Farmers, and Foodies Are Healing the Soil to Save the Planet."

"Plants remove carbon dioxide from the air and, combined with sunlight, convert it to carbon sugars that the plant uses for energy. Not all the carbon is consumed by the plants. Some is stored in the soil as humus... a stable network of carbon molecules that can remain in the soil for centuries. There in the soil, carbon confers many benefits. It makes the soil more fertile. It gives the soil a cakelike texture, structured with tiny air pockets. Soils rich in carbon buffer against both drought and flood." – Kristin Ohlson

Careful tending of the land, backyard compost heaps included, can put some of the carbon emissions caused by other human activity back into living ecosystems. I am a carbon farmer as much as I am a suburban dad who keeps a nice lawn for the kids to play on and a garden with fresh organic vegetables to put on the table. My pile is my back forty. Its yearly crop of rich, dark humus, teeming with life and recycled nutrients—chief among them repurposed carbon—is what makes my garden and lawn so prolific in the six months of the year when this part of the world is green and growing.

In addition to potentially saving the world, my hobby offers me plenty of outdoor exercise. It keeps me at home and out of trouble further afield. And it costs next to nothing. If this all sounds simple and skinflint, know that the payoff is profoundly rich and complex. The way I see it, my artisanal efforts to supply my pile with a wide-ranging buffet of ingredients give it uncommon potential. Then again, I imagine most every composter believes, like the parents of Lake Wobegon, their heap is above average. But after every work session, as much as I tell myself that this is no garden-variety pile I'm making, it always ends up looking the same—a big ol' heap of dead leaves.

Until, that is, the first hard freeze, which reveals the unmistakable sign that my backyard bio-engineering is paying off—on frosty mornings a vent of steam rises from the crown like the plume of a brewing volcano. I won't much disturb this big fat brown cocoon until early spring—only scuffing up the top every week or so to bury a supply of fresh, hot compostables within it. My pile cooks under its own weight and energy. It's taken everything that I can throw into it and will now make that lot its own.

All the Trimmings

The Latin origin of compost is *compositum*, which can mean a few things, among them "made up of little pieces." The word *compost* as we know it today, "a mixture of leaves, manure, etc., for fertilizing land," stems from an Old French word, *composte*, the roots of which are "bring" and "together." My backyard compost heap comprises all that and more, both in and of itself, and in the way it brings together the small community of my nearest neighbors. There are four families whose homes surround my corner lot. I've come to rely on them to help raise my pile, and with them I gladly share the final products of our homegrown bounty. In their own ways, my neighbors value my hobby as a convenient depot for their own garden and kitchen waste as well as a source of potting soil and topdressing for their own gardens.

Danute and Michel Tremblay have raised their close-knit family of four daughters in a small house on the southern side of my property. They keep their kitchen scraps in a lidded ashcan by the back door for me to collect on a weekly basis. A frugal family that rarely goes out to eat, their kitchen contributions to my pile far outstrip that of my own. The girls also have a pet rabbit, and from its hutch I get a steady supply of dank litter and dried pellets. Manure is about the most potent source of nitrogen you can add,

and the end product of a caged hare is more convenient to work with than, say, what comes out of the back end of a cow.

My neighbors to the west are the Rosens, who are ever friendly when talking over the short chain-link fence that separates us. For these super seniors I haul away most of the leaves of the sycamore tree on my side of the fence that collect in messy drifts in their driveway. You could argue (and I think they do over the dinner table) that all those leaves are my responsibility anyway. I don't mind. Besides, sycamore leaves make particularly good leaf mold. My pile is a mix of altruism and self-interest, which is another way to define community.

The Grissoms, across the main street to the north, donate grass clippings all summer long—Dad Carl has standing orders to keep his two adorable young daughters from tracking fresh grass clippings inside onto Mom Sarah's plush white living-room carpet.

Finally, across the side street to the east live the Favreaus. Pierre is a French Canadian who married Joanna, who grew up in the house where they raised their family of three children. A retired contractor with an eye for detail, Pierre used to spend hours sucking up every last leaf from their small, tidy lawn with a vacuum attachment to his electric leaf blower. It would take him an afternoon to fill a garbage can or two with pureed mulch, which he would then load into his pickup truck to haul off to the yard-waste center. We're friendly, even if I can still hardly understand a word of his Quebecois patois after all these years, and I am only too happy to take that bin of finely chopped leaves off his hands. Most prized are the contributions from the two large Japanese maple trees in

his yard, which drop their vibrant crimson leaves late each fall. I covet those delicate, star-shaped leaves like none other, and over the past several years have gladly raked them up to sweep onto my pile as red-velvet icing on the cake.

Other neighbors come into play on a sporadic but always welcome basis. Several times a year I trundle the Toro a couple houses down the street to mow the small front lawn of an elderly gentleman. His grown children live out of state and his only company is a home health aide who stays with him during the day. Dementia has robbed him of conversation, but he always had a friendly wave for my son and me as we passed by. Now he watches from inside the glass outer door as I mow, smiling below the oxygen tubes hooked to his nose. I'm fascinated by his grass, a thick-bladed variety unlike anything I've seen around these parts. I wish I could ask him about his lawn, as it grows slowly and is so thick I've yet to see a weed on it.

Across the road from the Favreaus is a single woman who lives with her elderly father. Perched on a slope of the rocky ridge that extends far down the street, her tidy house with its white picket fence is lorded over by several large oak trees, the leaves of which fall on the street side of her fence. I still have the thank-you card she kindly sent me after my down-the-street friend Don and I swept her leaves onto my bedsheet and dragged them away. The card reads, "Anyone can be cool, but awesome takes practice."

Since the late 1980s, Connecticut towns have been required to recycle a number of things, including all leaves. Most of the yard waste is collected at a town-run facility; it's where I bring any branches and rubbish too ungainly to compost. I've heard that the town's yard waste used to be composted locally, but some years back the local recycling operation was blocked by neighborhood opposition to the thought of mold spores. Though nearly denuded of trees two centuries ago, southern New England is now heavily wooded. I can't imagine how you'd make the argument that a local compost yard would produce any kind of mold that's different from what comes out of all the woods, bogs, fields, and yards that make up our corner of the world, but that's NIMBY for you. So the town shut down the composting facility and sold off its machinery.

In *Nature's Best Hope*, Douglas W. Tallamy advocates for a new way of imagining the patchwork of private domains that characterize modern suburbia:

> "Your property abuts your neighbor's property, which abuts another property, and so on. It is more accurate to envision your property as one small piece of a giant puzzle, which, when assembled, has the potential to form a beautiful ecological picture." – Douglas W. Tallamy

In our neighborhood, composting is very much an IMBY affair. Aside from generating lots of nice neighborly feelings, I figure I take in the bulk of leaves from a nearly three-acre reach of suburbia, counting pavement. That's six homes that have largely gone off the grid of the town's fall leaf cleanup.

It's Thanksgiving. My pile is freshly stuffed with a new supply of seaweed gleaned from the local beach after a pre-feast walk. My kitchen is abuzz with visitors—the Tremblays use my oven for the turkey that is too large to fit in their stove; the Favreaus stash their prepped casseroles in my freezer. Both are generous about returning the favor with plates of pies and other Thanksgiving dishes. The leftovers of the leftovers find their way into my pile with a final gathering of autumn leaves.

Endless Fodder

It's been a delightful fall; no early nor'easters, just a steady drop in temperature and daylight. The leaves have fluttered to the ground according to their own schedule and the occasional gust of autumn wind. But compost heaps don't thrive on leaves alone—at least not mine. True, a pile of leaves will, in time—a year or two or three in these parts—turn into leaf mold, a very good and basic kind of humus, to be sure. But where's the fun, the art, in that? So along with every fresh deposit of leaves, I try to add a sweetener, a catalyst of organic energy to inspire this particular heap to be something special.

There are four full seasons to this land at its privileged latitude, and my pile continually stimulates my backyard and me as a bio lab of a project in which one year's bounty turns into the next year's source of continued vitality. Each fall I gather the freshly dead ingredients of the growing season. Over winter and into spring, the collection melds and molds and is amended into something new. Late each summer, I spread loads and loads of wheelbarrows full of finished compost across the garden and lawn, which, in time, return a bumper crop of harvested life back into my pile. A compost heap is wonderfully useful to a suburban gardener. It represents recycling in its most elemental, home-brewed form—a hands-on effort that sustains both the garden and the gardener year-round. And in

this age of disconnectedness and drift, one of the chief marvels of tending a compost pile throughout the year is how it realigns me with the seasonal pace and cyclical progression of nature.

"There is something infinitely healing in the repeated refrains of nature—the assurance that dawn comes after night, and spring after winter." – Rachel Carson

Set in the corner of a small but lush backyard in the verdant, woody suburbs of Fairfield County, Connecticut, the heap is well situated. By and large, what my land and life produce and consume stays on the property. As a bonus, it reaps the rewards of the nearby seashore and local farms and stables. My pile weathers the swings in temperature, feeding off its own inner resources over the long, cold months when everything else is dormant, then drawing strength and sustenance from the spring and summer sun.

Nearly fifty inches of precipitation, on average, falls on this landscape each year. The regular dousings of water are as vital as air. Sometimes it's a gully washer of a thunderstorm, other times it's a foot-thick covering of snow. Generally, though, this essential ingredient comes throughout the year in near-weekly doses of an inch or so of soaking rain.

Making compost is an act of creation through destruction. Part plant, part animal, the domain of the multitude of microorganisms

that make up the preponderance of life as we know it, my pile is an endlessly looping live performance that plays out through the seasons. The heap is more than the sum of its parts. It's a process that produces something that scientists are still trying to describe, let alone precisely define. Compost is not dirt, exactly, nor fertilizer. "Soil amendment" hardly covers what scientists call *humus*, Latin for "earth" or "ground." Wikipedia sums up humus as "the fraction of soil organic matter that is amorphous and without the cellular structure characteristic of plants, micro-organisms or animals." How do you define something that by definition is amorphous? Wiki takes another stab: "In agriculture, humus is sometimes also used to describe manure, or natural compost extracted from a forest or other spontaneous source for use to amend soil."

If you can't exactly define something, the next best thing is to describe it, and in that regard, my compost pile is truly a little bit of everything. There's no exact recipe for making humus, and managing a compost heap is like being a chef who doesn't have to follow a recipe to cook up a finished product that somehow always seems just right.

What a compost pile is—and ends up being—depends on where it's made and what it's made from. This weekend, I dumpster-dove again for coffee grounds because the seaweed I usually gather from our local beach is in short supply. The moist, crumbly granules of spent coffee grounds have a microporous structure like charcoal and contain a wide range of useful microorganisms. "Spent coffee grounds are from a fruit nut that has been ground and had boiling hot water poured through it," explains composter Mary Tynes. "It isn't medical waste, or something that has been in someone's

mouth. It is the cleanest garbage around." Problem is, she adds, most people are too embarrassed to ask for someone else's garbage. Not me.

The last time I was inside my local java hut and asked the barista behind the counter if she had any coffee grounds I could take home with me, she had trouble processing the out-of-the-ordinary morning order. It's not easy to explain the intricacies of composting when there are three edgy, undercaffeinated people in line behind you—hence the dumpster diving. But some coffee shops around the country are making it easier to recycle and reduce waste. Starbucks, for instance, offers spent coffee grounds to composters, free for the taking, through its Grounds for the Garden program.

Each fall I construct my pile like I'm making lasagna, with as many layers as time and temperature allow. The pasta is the fall leaves, and in between is all the good stuff: crumbly coffee grounds, gloppy seaweed and gangly salt marsh hay, copious grass clippings, dollops of green manure from horse stall or rabbit hutch, endless gleanings from garden and kitchen, and whatever other spoonfuls of organic secret sauce I can find. In it all goes, layer upon layer, ingredient after ingredient, to simmer and cook. Add water, air, and the sweat of tossing and turning at regular intervals, then allow for countless living things, seen and unseen, to feast and fight and die within.

DECEMBER

The Best Fertilizer in the World

I'm blessed to live within easy reach of the ocean, and it's to the beach I go to bulk up on the greenest of greens for my pile—seaweed.

This is not a new idea in these parts, as I discovered some years back at the local historical society. The exhibit there detailed the area's agricultural roots, beginning in the 1830s, which developed richly with the "successful maritime exportation of fish and produce to New York, Boston, and beyond. By the Civil War, Westport was the leading onion supplier to the Union army, and onion farmers used nutrient-rich seaweed as fertilizer."

As I drive along the narrow, winding road my house sits on, I pass by two old onion barns. The smaller was long ago converted into a house; the larger, a two-story wood structure, is now used as what looks like a pool house for the renovated home in front of it.

My neighborhood is just a mile or so away from several public beaches strung along the northern shore of Long Island Sound. Connecticut is a state of rocky coves, sandy beaches, and tidal-river marshland. Its many estuaries are among the most ecologically

important in the country. I drive to one of the local beaches often in the fall, with Miller, my rescue mutt, sniffing sea breezes out the side window and a big plastic bucket in the back. Depending on the season, the weather, and the wind, high tide usually leaves a scraggly line of flotsam. Most of it is a motley salad of different kinds of seaweed and rotted reeds of salt marsh grass. The wrack line, they call it. It's always a good day when you are at the beach, and on most visits within an hour or so I can tire Miller out and fill up my keg-sized bucket with thirty or forty pounds of fresh, ripe seaweed or—just as well—a lighter mix of salt marsh hay.

Today's catch is good; a recent storm has pushed up a dense patch of detritus along a rock jetty close to the parking lot. The seaweed is yellow and brown and green and chopped by the waves into small, mushy pieces. The layer I set upon is a half-foot deep and flecked with all kinds of seaborne slop, a Sargasso Sea at my feet. I turn the plastic tub on its side and scrape up the briny mix with a three-tined hand hoe. Caught in the tidal ebb and flow are dismembered crab legs and molted carapaces of baby horseshoe crabs. Shells of mussels, clams, and oysters dot the mix, and in they go, too. The clattering seashells, which slowly break down into their basic components of lime and calcium, offset the acidic mulch of all the leaves. I've also heard that seashells give tomatoes more flavor, so I flick stray shells from the seashore straight into the vegetable garden. I always have to separate out a few bits of Styrofoam, plastic, or other inorganic debris—a bottle cap, snags of fishing line, a deflated Mylar party balloon with string. I love bringing this bit of the beach back home with me. The bucket smells like part wet swimsuit, part low tide, and all pure summer.

Mike McGrath writes in his *Book of Compost*, "Seaweed contains trace elements, micro-nutrients and plant growth compounds you'll never find in any chemical fertilizer—or even in most organic ones. One study found that seaweed-fed plants produced a third more tomatoes than non-seaweeded plants; and in another study, seaweed increased strawberry yields an astounding 133 percent."

"Talk about magic seaweed," notes journalist David Kirby in an article published on takepart.com, wonderfully titled, "How to Stop Farts From Warming the Planet: Feed Cows Seaweed." Though I worry about the amount of methane, however negligible it may be in the greater scheme of things, my puny pile may fart out, I take comfort in knowing that the seaweed I load into it may also be a solution to a far greater problem. "A single type of seaweed could cut greenhouse gas emissions, fight ocean acidification, remove invasive species, restore fisheries, and help coastal economies around the world," Kirby writes. Evidently, the seaweed contains a compound that helps disrupt enzymes used by gut bacteria in cows to produce methane, which has up to thirty-six times the global warming potential of carbon dioxide.

We may be only just now finding out how basically good seaweed is, in situ and in agriculture. Garden writer Eleanor Perenyi, in *Green Thoughts*, her classic account of gardening along the Connecticut coast in Stonington, also sought out seaweed, which required hiring "a man with a pickup truck and the willingness to scramble over wet rocks wielding a pitchfork, not a combination I find every day." Here in Westport, trade in the truck for an SUV, and that man is me.

"Like compost [seaweed] is a fertilizer as well as a soil conditioner, one of the oldest known to man. All marine peoples have used it. In seventeenth-century France, royal regulations established the kinds to be gathered and how they were to be used. It has twice the potash content of barnyard manure, making it perfect for beets, potatoes and cabbages, the potash lovers. More than that, it has the power to unlock minerals in the soil; it contains growth-inducing hormones that will increase the yields of tomatoes, corn and peppers. Plants given seaweed are better able to endure a light frost, and some are made more resistant to insect and disease attack. With those remarkable properties (some of which, it is true, have only lately been established by research), and given the high cost of commercial fertilizers and pesticides, you might expect to see the gardening citizenry of both coasts swarming over the rocks and beaches. You don't, partly because no high-level interest exists to care to tell us."
– Eleanor Perenyi

Another inspiration for adding seaweed is *The Field*, a fine if unsettling film by Jim Sheridan starring Richard Harris, John Hurt, Sean Bean, and Tom Berenger as the rich, handsome Ugly American. The title role is played, with convincing Irish charm, by an acre of lush green pasture enclosed by a rim of ancient stone walls. Bull, played by Richard Harris, has tended the rented vale his entire life,

turning it from barren ground to most productive pasturage, where he raises fresh hay and straw to feed his livestock for market.

The movie begins with Harris and Bean, as his mulish son, collecting heaping strands of giant kelp fronds from a rocky beach, packing the lot into wide-mouth wicker baskets on their backs and schlepping the harvest over hill and dale back home. Their arduous trek plays out wordlessly over the opening titles. Cresting the last slope between the sea and the field, Harris plops down his basket. Gazing over the valley, he says to his son, "God made the world, and seaweed make that field, boy."

It's a tragic film, and near the end, old Bull tells the American, who wants to buy the land out from under him, "It's my field. It's my child. I nursed it, I nourished it, I saw to its every want..."

If not to the same morbid end, I feel the same way about my pile.

Clean Margins

I am getting ready to go to work when a cherry-picker and a covered dump truck hauling a chipper pull into the driveway of my across-the-street neighbor, one house down, opposite the Rosens. The tree crew is there to remove an old white pine that overhangs one of the two houses on the property. Over the past several years, this neighbor has reaped what he sowed long ago. Walter is a widower who has lived in the same spot for more than five decades. Some years ago, once his children were grown, he built a second house on the property, an A-frame for himself and his wife, and rented out the original home. For privacy, he planted a row of hemlocks along the driveway that separates the two houses and also placed pine trees in front of the main house. Behind both houses is a ridge studded with aged oaks that grow from crevasses in the granite ledge. His property, once a shady, private haven, has become an aerial minefield. Two nor'easters ago, a fallen oak took out his sailboat and deck, as well as a portion of the A-frame roof. The fir trees are tall and spindly, long past their prime, as is the remaining white pine.

Like many homeowners in the suburban Northeast, I face a similar problem, a double-edged sword of Damocles hanging over my house. The tall trees on my property are stately and provide much shade and, of course, fuel for my pile. But they are also a hazard.

Over the past few months, the Rosens next door have asked me to do something about the sycamore that rises from the border of my yard. It spreads its long, craggy branches over the cars in their driveway and the utility wires that stretch from the telephone pole at the street to the corner of their house. I'm sure they'd be happy if I took the whole tree down, but it's my favorite on the property, majestic even. On clear dawns, the rising sun touches the highest branches first, then spreads its rosy glow down through the sycamore's dappled limbs and trunk. It looks as if the gloom of night is draining back into the ground.

As the tree crew sets up shop at Walter's, I ask the foreman to quote me a price on trimming the sycamore. Almost as an afterthought, I ask him how much it would cost to take down the last remaining swamp maple in the backyard. Its branches cast shade over the largest stretch of lawn and its roots infiltrate the garden beds, and, no doubt, the base of my pile.

I add that I'll take the logs from my yard for firewood, and the trimmings as wood-chip mulch to spread across the perennial beds. His price is so reasonable I agree to the terms on the spot, and he says he'll be back tomorrow morning to do the job. Good for me, good for him. It's a no-brainer to unload a truck one door down the street rather than haul it thirty minutes away to pay some rural landfill to offload it.

I take the day off, a Friday, as a busman's holiday. What you would need to pay me to climb a tall, swaying tree with a chainsaw far exceeds what I'm paying today. They begin at 8 a.m., backing the industrial-sized wood chipper attached to an enclosed dump truck

onto the lawn to start with the maple. The double-axle tracks a set of indentations that I can only hope the turf will recover from eventually.

Within an hour, the arborist has worked his way from far up the canopy down the main trunk, sawing off man-sized sections as he goes, which topple to the ground with heavy, pounding thuds. Another worker slices the logs to firewood length, and I help stack them along the backside of the tool shed. As another of the crew feeds the remaining branches into the chipper, I rake up the thick layer of sawdust from around the trunk and drag the minced shavings over to my pile. Sawdust has an extraordinarily high carbon to nitrogen ratio. Your garden-variety compost heap clocks in at a 25:1 or 30:1 ratio; sawdust is 500:1, so a little goes a long way. But like coffee grounds, its raw, granular nature makes it easy to add to the mix, even if its raw carbon must be as tough as diamond dust for microbial critters to chew on. Besides, I like abiding by the guiding philosophy of what grows on my property, stays on my property.

As the crew moves the articulated lift to trim the sycamore, I spread a large plastic tarp across my driveway for the foreman to dump the chips onto it. Before I spread the coming mulch throughout the garden beds, I need to do some prep work. It's been two seasons since I added wood chips to the perennial gardens that ring the lawn; last year I spread loads of finished compost atop them. Since then, the border between lawn and garden has become a ragged line. The ryegrass, clover, and viny wild strawberry have infiltrated the flowers and ferns. I take the wheelbarrow and the flat shovel from its perch in the shed. Time for some border control.

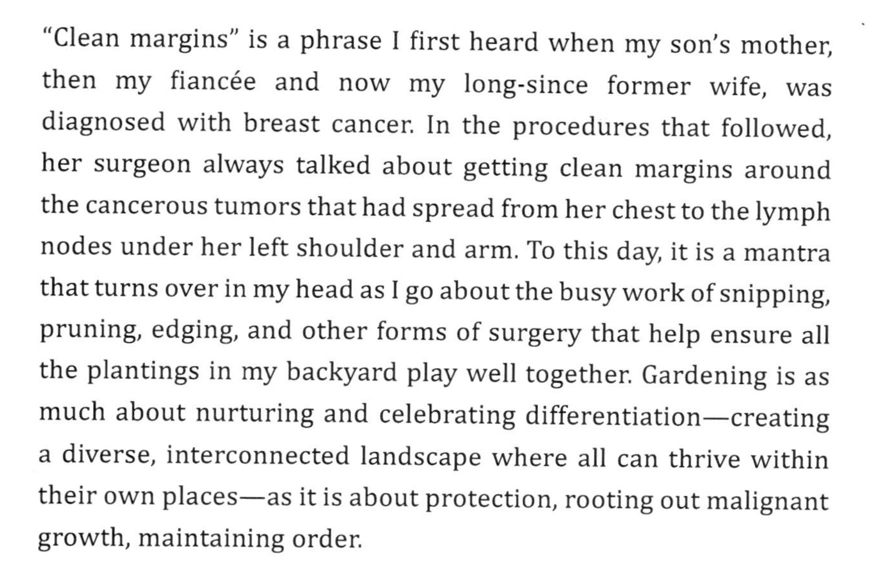

"Clean margins" is a phrase I first heard when my son's mother, then my fiancée and now my long-since former wife, was diagnosed with breast cancer. In the procedures that followed, her surgeon always talked about getting clean margins around the cancerous tumors that had spread from her chest to the lymph nodes under her left shoulder and arm. To this day, it is a mantra that turns over in my head as I go about the busy work of snipping, pruning, edging, and other forms of surgery that help ensure all the plantings in my backyard play well together. Gardening is as much about nurturing and celebrating differentiation—creating a diverse, interconnected landscape where all can thrive within their own places—as it is about protection, rooting out malignant growth, maintaining order.

I take a break from the tedious task of relining the garden border and drag the small tarp of sod I've culled. Each time I tidy up the edge line I remove a foot-wide strip of turf, expanding the garden bed and shrinking the lawn. All kinds of living things and decomposers dwell in the root zone, and these clumps of dirt topped with grass will introduce them to my pile.

On my way back from the heap I notice new clusters of a glossy green weed sprouting here and there among the now-brown grass. They are offshoots of a bigger patch I recently identified as lesser celandine. For years I've tangled with this weed, which I've learned is a highly invasive import from overseas. It's a type of buttercup, with small yellow flowers that are among the first to bloom each spring. I've tried to contain it by hand weeding with the dandelion digger to no avail. Not only does celandine spread by seed, but it also has bulblets attached to the stem and tubers that spread

further through the soil. Miss a little bit and you only encourage the rest of the low-growing ephemeral to expand into the disturbed soil around it.

I make a detour to the shed to grab the spade and with the pointy end excise the mother clump wholesale and dig out the newer infestations around it. Left unchecked, celandine will grow in dense mats, creating a monoculture that crowds out all other plants, with no native insects or microbes to stop it. I dig deep with the spade, trying to remove all bulblets and fat roots, then fill in the holes with the choicest sod from the garden border. Clean margins, to be sure. I tamp the new turf in place by foot and stash the celandine clods in an old paper bag I'd emptied of leaves from across the street last fall. No way are those nutlets going in my pile.

The tree crew is finished by lunchtime. The Rosens come out to review the work and are greatly relieved to see the limbs that once spread over their driveway have been whacked back to near the property line. Clean margins.

As we talk, the boss man backs the wood-paneled truck onto my driveway. I didn't think to check inside before requesting that it be emptied. Still full from the job across the street the day before, the back bed tilts to disgorge a truly daunting amount of wood chips. I watch slack-jawed as they cascade over the tarp and onto the grass on either side of the driveway. The crown of the pile is nearly head high. You could hide a car under it.

I look at all the chipped-up wood and know now that I will have a busy weekend. Over the years, I've spread truckload after truckload of the remains of local trees and bushes across my property. If it's

not grass or patio or pavement, the ground I tend is covered by a layer of what tony landscapers call arborist mulch. Wood chips aren't quite as virtuous as compost, ecologically, but they do have their benefits. A three- or four-inch layer of chips spread among my garden beds prevents weeds from sprouting. It soaks up rainfall and slowly releases it, reducing the need to water the flowers, bushes, and shrubs that ring the perimeter of my property and surround the house.

A blanket of freshly minced trees also gives the ground around the perennials and underneath the trees a uniform, tidy appearance, though for a few weeks my yard has the look of the kid with the new sneakers at school. And for a time, especially after it's wetted by the first rain, a fresh layer of wood chips gives my yard a Christmassy, piney scent. When I happen upon a load made of black birch, I catch a whiff of Wrigley's wintergreen.

Some houses in my neighborhood are landscaped with store-bought mulch made from coconut husks or nutshells or ground-up bark processed from who knows where—often dyed an unnatural shade of red or mocha brown. I prefer to get my wood chips unvarnished and free of charge, then spread them myself. I also like knowing where the chips come from—a mystery load of chopped-up tree or brush could introduce some blight or bug into my landscape.

"In many urban areas, arborist wood chips are available for free, representing one of the most economically practical choices," says Linda Chalker-Scott, associate professor and urban horticulturist at Washington State University Puyallup. "Unlike the uniform nature of sawdust and bark mulches, wood chips include bark, wood, and often leaves. Additionally, the materials vary in their size and decomposition rate, creating a more diverse environment that is subsequently colonized by a diverse soil biota. A biologically diverse soil biota is more resistant to environmental disturbance and will in turn support a diverse and healthy plant population."

Over a season or so, wood chips spread across bare ground are broken down by rot and fungi and earthworms to become a deep new layer of biomass. As I marvel at how all that minced wood reduces to so little, I realize that what I'm shoveling is mostly water and air, bound only for a moment into bite-size pieces of carbon.

"The greenness and fertility of my garden are due to vast quantities of mulch, everything from compost to salt hay to seaweed. It is an organic substance whose benefits extend to the soil itself, improving its structure and enriching its fertility to the point where it needs nothing else. An organically mulched vegetable garden never requires tilling, digging or hoeing, and is scarcely weeded." – Eleanor Perenyi

After Saturday morning errands, I layer up and haul the wheelbarrow from behind the tool shed and load into it a set of tools—the wide-tined hay pitchfork, two rakes, and the wide-brim shovel. I wheel them to the front of the yard and set up shop beside the mound of tree mulch, already steaming with the raw, aromatic

scent of sap and fermenting wood pulp. I feel like a rube for accepting the chips without first peering into the truck to see what it contained. I didn't fall off the turnip truck; it fell on me. I eyeball the mound and try to mentally parcel it out, then glance up and down the street and wonder on which of my neighbors I can pawn off the excess chips I fear will remain after I've covered every square inch of my available ground.

I lean the wheelbarrow against the steepest side of the mound and plunge a hay pitchfork into the peak, using gravity to fill the basket with several scraping thrusts. Strange how striving for a low-maintenance garden involves so much work. Before long, I've peeled down to just a heavy work shirt as I trundle the wheelbarrow back and forth across the firm, cold ground. The need to dispense with a massive pile of anything is always a daunting task. But volume measured in cubic yards becomes manageable when spread across a lot of linear square feet. Call it sheet composting. The load of wood chips is a pop-up version of my pile; here today, gone tomorrow.

Spreading wood chip mulch is both a physical and mental exercise. Most of the energy expended is in the transport. I schlep a dozen or so wheelbarrow loads to spots across a section of garden before stopping to rake out the piles, taking care not to bury an azalea here, there a phlox. This batch of chips, I'm relieved to find, is a fluffy, airy mix of fresh tree potpourri, suffused with shredded evergreen boughs and diced sycamore balls, nearly as light as cotton candy. I spread it more thickly than usual, without making life too difficult for the daffodil and crocus bulbs buried beneath.

I enjoy wielding a wheelbarrow and the business end of a pitchfork and rake. It's a weekend warrior's workout, and by late Sunday afternoon, I've packed up the tarp and fired up the leaf blower to scoot the remaining flecks of chips from the gravel driveway. The garden beds are now cloaked in a mesh of fresh-chopped mulch. My pile has benefited as well, for at the bottom of all those chips was a pocket of ground-up leaves, masticated sycamore seedballs, and sawdust shavings. I add two wheelbarrows of the fluff straight onto my pile. I'd salute myself in triumph for a backyard gardener's job well done, if I wasn't too tired to lift my arms.

Breathing Room

It's a balmy Sunday in mid-December, and although the neighbors are busy decking out their homes and yards with Christmas decorations, it's unseasonably warm. With a record high of near seventy degrees today, I leave my own twinkly lights in their boxes in the attic. Instead, I will devote the short winter day to sprucing up the backyard and taking care of outdoor chores.

Being a bachelor dad with no local family, I have always tried my best to make the holidays special for my son. One father-son tradition, which we accomplished this morning, involves heading a couple miles down the road to a nearby wildlife preserve, managed by the local Audubon Society. Consisting of two parcels on either side of a street leading to the beach, thirty-six acres in all, the sanctuary is in the midst of a neighborhood-led restoration to rid the open space of decades of tick-infested invasive growth. The goal is to create a mix of coastal woodland and meadow with native plantings for pollinators and grassland birds.

The property is home to a Christmas tree farm, a relic of days past when the area was still rich in working farms, not just Gold Coast real estate. Each holiday season, the old farm turned conservation area hosts a cut-your-own-tree operation that helps fund its future as restored habitat. It's also an opportunity to give a young boy

a lesson in making decisions when options abound. Which tree would speak to him? We wandered endlessly through the aisles of the grove, assessing each tree for its own character and prospects of fitting inside our tiny home.

Another holiday tradition we enjoyed some years ago started as a family joke. Our living room is so small, I recall musing to Cole, that maybe we should have a tree that's wide at the ceiling and narrow at the floor. That way we wouldn't have to move the sofa to watch TV. Besides, I teased him, wouldn't that mean Santa would have more room to leave presents? Perhaps the notion of an upside-down tree hanging from the ceiling had come to me the Christmas before, when we adopted Miller as a gift from his grandmother and he chased the cat up into the tree, with predictable results.

In any event, the next season, as we set out for the tree farm, Cole said to me, "We're getting the upside-down tree this year, right, Dad? 'Cause I kinda told people at school we were doing it..."

As I was already a total embarrassment to my then middle-school son, that settled it. "For sure," I said.

I'll leave out the particulars, though I will mention they involved hole drilling, toggle bolts, eye hooks, and an as-yet-to-be-patented water-drip system that mostly watered the floor. But the tree went up, or down, I should say—and it was a delight. For one, it made searching for just the right tree a lot easier; the scragglier the better. And it turns out an upside-down Christmas tree is surprisingly easy to decorate. You can spin it, which makes stringing the lights a breeze, and the ornaments dangle nicely from under the bent branches. In truth, it was like one big ornament itself.

We enjoyed our inverted tree immensely, especially after one of Cole's school chums came over for a play date and asked very earnestly, "So, how are you going to get the presents to stick on the ceiling?"

That was then. Meanwhile, this year's straight-up tree waits on the porch and my pile abides, Buddha-like, in the back of the yard. Over the past two months, I've made a dozen or so leaf roundups, topping off each new load with layers of grass clippings, seaweed, kitchen scraps, rabbit litter, coffee grounds, and whatever other ready-to-rot organic filler I come across.

After all those additions, my pile has assumed its usual winter repose. Its bulk seems permanent, immutable save for the tendrils of steamy vapor that waft up from its mounded summit on cold mornings. But I know that underneath its mantle of damp brown leaves and seaweed flecks, it is stewing. Left to its own devices, it will slowly cook through winter, a crockpot of compost. Despite the warning about a watched pot, I can't resist tinkering with it. Over winter, I'll continue with my regular deposits of kitchen scraps and such, carving out holes for the slop, then covering the fresh material with old leaves. I like to leave a rim of undisturbed leaves against the back fence that contains the heap. It's for insulation mostly, but I also don't want any banana peels or eggshells hanging out for critters or me to see. My pile keeps its secrets well.

Since fully turning the heap will have to wait until spring—after the mass has had all winter to cook down into a more manageable size—the most useful thing I can add now is air. Today I plan to give the heap a good prodding, using an eight-foot length of rebar

I snagged a couple years ago from the Tremblays' scrap bin. The rusty stick of half-inch steel weighs just a couple pounds, but climbing atop the log walls and thrusting it a couple dozen times into the midst of my pile gives me a good, quick workout. As an aerating tool, it does a fine job of making my pile more porous, for both air and the water to follow. I've read that neither air nor water seep much more than two feet into a pile in any direction, so why not lend a helping hand, even if it means poking a slumbering bear?

"What you are doing when you construct an aerobic (with air) compost heap is creating the right environment for the billions of microorganisms that make the compost happen," counsels British garden writer Nicky Scott.

"A happy heap will have a balance of air and water, just like a squeezed-out sponge; the whole surface area is coated in water but there are air spaces in between. If the pile is too dense, squeezing out all the air, then all the beneficial life forms in the compost heap are not going to survive and will be replaced by the 'bad' microbes–the anaerobic (without air) ones that are responsible for all the bad odours you get from putrefying substances. This is bad news for your compost and if you put this material on your plants it can be toxic to them. So creating air spaces in the compost is vital." – Nicky Scott

Tumbling a compost pile is the surefire way to add gulps of air, but mine is too big and unwieldy at this point in its life cycle. Instead, I try to give it good lungs from the start, building it up in layers and using plenty of porous material along the way. Those materials consist mostly of the hollow reeds of salt marsh hay gathered from the seashore, but also armfuls of spent tomato vines and flower stalks from the cutting garden. Some years, the deer allow me to grow sunflowers along the fence bordering the vegetable garden, and, come fall, their thick, fibrous stalks make excellent ventilation shafts. The soft centers rot out, creating hollow tubes through which both air and water flow. Some compost experts advise starting a heap by first sticking a length of perforated PVC pipe in the middle, or a rolled-up tube of wire-mesh fence, to serve as a more permanent chimney. I may try that trick someday, but this season I'll stick with sunflower culms augmented by periodic jousts of slender steel.

The rod thrums like a tuning fork through a section of dried leaves, its ribbed rebar providing sensory feedback through the vibrations in my gloved hands. When the blunt end strikes an oyster shell from a seaweed haul, it tings like sonar. I hazard a guess at a tough but squishy part. The jack-o'-lantern tossed in after Halloween? I punch on, and when the tip meets hard ground, I feel the jarring endnote all the way up to my elbows. Sometimes as I pull the rod back up and out, my pile wins the tug of war, and the pole slips through my gloves. I grab tight and pull harder, fearing that one of these days I might stick the shiv right up under my chin. I skewer a dozen or more times from each side and front and back, varying the angle of entry each time, turning my pile into a pin cushion. In my mind's eye I see each slender tube of air

as a superhighway where countless unseen bugs and bacteria will mix and mingle.

The end of the rebar glistens with wet steam. I pull my glove off to confirm: the metal is warm to the touch. I wonder if the heat is caused by friction alone, so I stick the piece of cooling steel back into the heart of my pile and wait a few beats before slowly drawing it back out. The bar practically sizzles. I take it as a good sign that I've created a churning, burning mix of earth, water, and air. It's smoking hot, my pile.

Marking Territory

The parameters of my pile are shaped partly by my ambition but mostly by the sheer number of leaf-producing trees within fetching distance. Its full measure is about eight feet wide, eight feet deep, and five, maybe six, feet high at peak. Which is good, because in these climes, a compost pile needs to be of sufficient size to sustain its own internal combustion.

Old hands and the research suggest that four cubic feet is the bare minimum for an outdoor, uncovered heap to keep the "hot" bacteria going. Anything smaller isn't really a compost heap—just a prospect of one, a mess of dead leaves on cold hard ground.

"In most areas of the continental United States, a compost pile needs quite a bit of mass to be self-insulating and maintain ideal temperatures," advises Stu Campbell in *Let It Rot!* "A pile that is too small may lose its heat so quickly each night that it will cool off, or even freeze, quite readily. Pathogenic organisms, weed seeds, and larvae will not be killed, slowing the whole process. If you want hot, fast compost, your pile should measure at least 1 cubic yard."

Piles bigger than average—like my own—require more physical effort to sustain the proper mix of air and water needed to fuel the decomposition process, but I don't mind. My pile provides me much

more than an ongoing harvest of compost. It gives me an excuse to get outside every so often, plenty of exercise, and a purpose. As my pile serves me so well, I heap it with produce and praise and shower it with attention... and pee.

As Charlton Heston once explained to Dear Abby, in response to the lady ("The Whiz-zard's Wife") who feared her husband's habit of urinating on their lawn was inappropriate: "So it may be, but the fact remains that all men pee outdoors."

It's true. One of the reasons I keep my pile as tall as I do is to be able to take a leak behind it without fear of pissing off any passing neighbors. Most mornings when I let the dog out, he pees on the front of my pile while I take the backside. After a cup or two of coffee for me and his morning bowl of chow, we may hit it again.

I'd probably pee outside even if I didn't keep a compost pile. I spent enough years in rain-deprived Southern California to value each toilet flush, and I've paid enough utility bills to know to conserve my supply of water for more important uses, like watering the tomatoes in July.

I also drink a lot of coffee and beer and, when home, try to spend as much time outdoors as indoors, often tromping around in muddy work shoes. Go back inside to take a leak? No. I piss on my pile. It's a convenient if somewhat furtive release, in a Huck-Finn, behind-the-woodshed kind of way.

To excuse myself further, let's hear more from Abby and company on the subject of "pee-cology":

DEAR ABBY: Marking our territory is only one reason for this age-old tradition. Boys have long enjoyed distance, accuracy and creative urinary competitions: knocking leaves off the trees in the fall, drawing pictures in the winter snow, protecting young fir trees from hungry deer in the spring, and dousing campfires in the summer months are just a few highlights.

Some may deride this as small-minded male nonsense, but on a global scale, this ritual has significant benefits to our environment. The flush water we save is substantial. At 2.5 gallons per flush, a man urinating outside just once a day will conserve almost 1,000 gallons of water a year. If one-fourth of the men in the United States saved one flush per day, we'd save more than 4.5 billion cubic feet of water per year.

If you consider all the rainfall that's channeled into storm sewers from our streets and parking lots, we're returning valuable moisture to the soil by urinating on our lawns. — NATIVE OREGON STREAMER, TILLAMOOK, ORE.

Standing at the backside of my pile, I do some calculating: A couple leaks a day adds up to more golden showers a year than I'd like to admit. But all that urine is good for my pile. It turns out this body waste that's (usually) flushed down the toilet can actually be recycled into a number of useful products. According to todayifoundout.com, "comprised of water, calcium, chloride, potassium, sodium, magnesium, urea, creatinine, nitrogen, uric acid, ammonium, sulphates and phosphates, urine's beneficial ingredients can be separated from its waste, and used to make fertilizer, medicine, brain cells and, yes, gunpowder."

The key ingredient is phosphate, which was identified in the 1660s by German alchemist Hennig Brand, who was trying to turn urine into gold. Instead, he turned 1,500 gallons of urine, likely collected from beer-drinking German soldiers, into what became a new kind of liquid gold. Under high heat, the phosphate in urine loses its oxygen and becomes phosphorus.

An essential element, phosphorus is the sixth most abundant element in any living organism. It's the "P" in the "N-P-K" fertilizer formulation, between "N" (nitrogen) and "K" (potassium). Phosphorus not only glows in the dark but can also be highly poisonous and combustible (white phosphorus is used in many destructive weapons, such as napalm). That said, human urine could hold the key to solving the climate crisis. According to *Mother Jones* magazine, "There's enough phosphorus in your annual output of urine to provide P for more than half of all the grain you consume in a year."

That may explain why in chronically famished North Korea, cut off from external sources of fertilizer and desperate to make its own supplies of manmade manure, workers are, according to *The Daily Beast,* "forced to bring two liters of urine per person per day to mix into the compost pile until the production goal is achieved."

The mandate, while draconian, is nothing new. Urine was considered such a valuable commodity in ancient Rome that the emperor Nero levied a urine tax. Laundries placed giant clay pots out in public for people to relieve themselves. Urine was used as a cleaning agent for washing clothes, for tanning leather, to bleach wool and linen, even for brushing teeth, I read on a website called

Ancient Origins. "The Romans believed that urine would make their teeth whiter and keep them from decaying so they used it as a mouthwash and mixed it with pumice to make toothpaste. In fact, urine was so effective that it was used in toothpastes and mouthwashes up until the 1700s." Nero's levy in the first century AD gave rise to the saying *pecunia non olet*, meaning, "money does not stink." A later emperor, Vespasian, used the urine tax to fund the construction of the Roman coliseum. Now that's what I call leaving your mark!

Early each spring as the grass begins to grow again, my lawn is pockmarked with bright green patches supercharged by both the dog and passing deer. I know from the splotches that these frequent doses of energy, in the form of nitrogen, urea, and other organic (and sterile) materials, are effective. Why not direct them into my pile? According to Nicky Scott, urine is "the cheapest and best activator to speed up the composting process. It adds nitrogen and water to woody, dry, carbon-rich material."

Some years ago, as I was beginning to take backyard composting more seriously, I went to the local garden store to shop for my pile. I'd read a little about "activators" that kick-start a pile's decomposition and wanted to make sure I wasn't missing some essential ingredient, an edge.

On the shelf I saw a squat bag of a powdery product called Bio-Excelerator, which promised "to offer the complete solution to generating rich, fertile humus—Nature's best soil conditioner." The package looked and hefted like a bag of flour; the backside label boasted that inside were "billions of microbes especially cultured

for composting. In addition to containing moderate and high-heat active microbes, special varieties are included that can speed the decomposition of difficult-to-compost organic matter. All are combined with special proprietary energy sources containing kelp and dried blood to ensure a rapid decomposition..."

Like a bottle of daily vitamins, it seemed to me, only more mysterious, in a secret-sauce sort of way. So I plunked down $11.99, plus tax. I shook the white powder onto my pile like so much pixie dust. I figured it couldn't hurt, and just might help.

Talk about pissing your money away. That was before I'd begun my winter reading of composting and gardening books and learned that my pile could do just as well left to its own devices, abetted by a surreptitious tinkle or two.

"Your nose may wrinkle at this, but human urine is one of the best additions for a nitrogen fix: it is entirely sterile, so it can't be harmful, and as well as containing a high percentage of nitrogen, it is also crammed with minerals and vitamins," concurs Claire Foster in *Compost*, another British title. "Many an owner of a vegetable patch (mostly male it has to be said) confesses to having the occasional pee on the compost heap—after dark of course—and it really does work wonders."

I take comfort in knowing my own end product can be so vital to the new beginnings that my pile is all about. Plus, I like having a good excuse to pee outdoors and the privacy that my fulsome pile offers in taking that relief. It's a momentary respite, allowing for the kind of sharp focus that comes with staring at the graffiti above a urinal or etchings on the backside of a restroom stall.

As I dawdle behind the heap, I see the seeds from the pumpkins smashed a week or so ago have sprouted through the sodden leaves, their tender pale green shoots striving to gain purchase. I look closer at a slender stalk of white; it's not a pumpkin sprout but a wing feather from some sort of waterfowl, probably gathered up with the batch of seagrass hay that now tops my pile like a frazzled toupée. The hollow nub of the quill is aswarm with roving creatures just large enough that I can detect their movement. They look like roly-polys only much smaller. I wonder whether these tiny scroungers came to my pile already aboard the molted feather or if they were resident scavengers with a taste for holiday goose. And what feeds upon these tiny mites when they are finished with the feather?

"Details, details, you might protest," writes paleontologist Richard Fortey in *The Wood for the Trees,* a biography of his own small patch of land in the shires of England.

> "I reply that the delight, as well as the devil, is in the details. To an animal of small size, particularly an insect in which the larva hides away discreetly to feed and grow, our wood is a *potpourri* of opportunities, quite a wonderland of niches." – Richard Fortey

My pile is a living curio cabinet, a menagerie composed of countless wild creatures I keep in untamed captivity. In return

for helping create, develop, nurture, and mature this creative act of deconstruction, I am rewarded in compound ways, not only in wheelbarrows full of compost at the end of a year, but also as a mental pit stop, every day.

I take a last look at the mop-top of blonde straw, which gives my pile a finished look, in a tousled, hayseed sort of way. I lean in to spot a baby clamshell dangling from the tip of a stem. The waning sun shines through the pearly skin; it twists in the breeze on a strand of sea green like a Christmas ornament.

JANUARY

Inner Workings

The first Sunday of the new year, a few days after the first hard freeze of winter. I consider my pile from behind the frosted glass panes of the kitchen door, squinting through the low morning sun for a sign of vapor rising from its top. What combustible forces, if any, are still acting within? How far into the heap has the cold seeped? What protection does the insulating cloak of leaves and seagrass straw provide? Have I given my pile the resources it needs to keep from shutting down completely?

I let Miller out and follow him into the backyard. Each step creates a crunchy footfall on the frozen grass. I take stock with a morning pee through the wire mesh that girdles the backside of the heap. A tendril of steam rises through the damp stalks of seagrass cross-hatched along the top.

A month ago, with the last big crush of swept-up leaves and gathered seaweed, my pile crowned higher than my head. Exhaling, it settles upon itself and is now chest high. As it assumes a more tranquil repose, it begins the year complete. A heap in full. Almost seems a shame to meddle with it.

But fuss I will, for I have a holiday's worth of gleanings from my kitchen and the neighbors to disperse. Plus, it's a mild winter day with a mix of rain and sleet on the way, and I could use some exercise and an outdoor diversion, a break between bowl games. This heap is my own private hot-stove league.

I do indoor chores while the sun slowly warms, first scooping the cold ash and charcoal bits from the fireplace, then turning to the overdue-for-cleaning, ten-gallon aquarium in the game room that's home to Bubbles, a red-eared slider turtle. My son and I rescued him from the house of a play date, where he lived in a Tupperware bowl on the kitchen counter. Our glass tank was vacant at the time. I'd had what I thought was the bright idea to let Cole's pet box turtle get some fresh air, so one morning took her out to the vegetable garden, figuring she could patrol for slugs and other pests: a guard turtle for the garden. When I got home that evening, the turtle had dug under the wire fence and was gone without a trace. You try explaining to a kid that his pet turtle ran away.

An illicit souvenir from a Chinatown shop, Bubbles was the size of a silver dollar back then. Ten years on, he's grown as big as a fist, with a personality to match. He'll eat anything I toss in the tank, from food sticks for reptiles to live crickets caught in the garden to the occasional serving of minnows from the pet store. The filter can hardly keep up. Bubbles is all piss and vinegar, and I love him for that.

Soon the turtle is paddling about in his tank of fresh clean water, and I have a bucket full of murky green turtle poo and flecked-off

shell scales to pour into the heap. Better to that end than flushing the slop down the toilet. Laced with nitrogen, urea, and who-knows-what other nutrients that make up a Chinatown turtle's night soil, these fetid gallons are like jet fuel for my pile.

I make the best use of this short day with a quick trip to the beach with the dog and bring home a half-barrel stuffed with salt marsh straw. Skirting the old wood stockade fence along the backside of my property, I check on the metal ashcan the Tremblays keep outside the back door to their kitchen and find Danute cramming in the last of the holiday leftovers. Her larder has been particularly bountiful of late, she tells me, because her second daughter, on hiatus from the local community college, is now working at an organic produce store. She brings home fruits and vegetables bruised or past their prime. The store's too pricey for my budget, but I'm happy to have access to these rich remnants.

I glance over Danute's shoulder toward the back of her small, fenced-in yard to notice a low ridgeline of leaves raked just far enough away from their picnic table to be out of the way. She is diligent in keeping her front yard tidy, and throughout fall has swept up her leaves into piles for me to drag away, but she has yet to tackle the backyard, as it's mostly the domain of her pack-rat husband, Michel—filled with stacks of salvaged wood, bins of pipe, and other takeaways from the town dump. She knows that on through winter I'll help finish the task, as my pile is like the bear in the storybook and always wants more.

After assembling my contributions at the base of the heap, I scrape its crown of seagrass hay to the side with the gravel rake,

then plunge the wide-tined pitchfork into the spongy, wet leaves, dragging forkfuls toward me to create a trench across the top. I dig deep into the time warp, down through the stratified layers of past heapings, freeing wafts of steam along the way. Two feet down, the pitchfork jabs into an impenetrable mat of pressed whole leaves—like a fork sticking a phone book. I reach my hand into the hole. The mat of leaves is cool to the touch. It seems my pile is combustible only to a certain degree. The cold of winter is winning out over the hot flush of organic fusion.

Into this newly formed trench I scatter bright white shreds of paper brought home from the office, then the ashcan full of the primo-produce scraps from next door, along with my own, lesser, kitchen wastes. After adding a layer of dried leaf scraps from the ragged front of the heap, the bin of seaweed and salt marsh hay is next. There must have been a mass molt among the crabs, for this batch is suffused with their carapaces. Bubbles would be pleased to see them marinating in his turtle soup, which drains through the freshly mixed mess.

Taking up the four corners of the bedsheet full of damp maple leaves from next door, I step up along the top of one of the log walls and unfurl the groaning load across the open maw. Over the next few months, I'll gouge out similar holes in a half-dozen places, hoping to spike the heap with enough hot spots to keep the biological processes churning through the cold winter. I finish by drawing the blanket of seagrass stems back across the top with the rake and sprinkle it all with a dusting of wood ash and charcoal crumbs from the fireplace. Newly stuffed with a fresh load of raw

organic material, my pile is once again whole. Underneath its new cloak of old leaves, it will continue the unseen magic of its transformation into something much less—and much more.

In *Improving Your Soil: A Practical Guide to Soil Management for the Serious Home Gardener*, Keith Reid estimates that "for every 100 pounds of fresh organic material added, a mere 1 to 2 pounds end up as humus." I'm fascinated by the disappearing act that is my pile. Its inner workings remain largely a mystery to me. Even soil scientists are still profoundly uncertain about what exactly takes place, biologically, underneath our feet. Michael Pollan and other close watchers liken soil to a frontier more unknown than the deep oceans or outer space. I consider composting more craft project than lab experiment and am happy to let my pile do its own thing, with a certain amount of input and creative direction. That said, both the art and science of making compost are well developed, and how-to advice is plentiful, whether from books or with a simple online search. I find it odd, though, that at least at my local bookshop, the home and garden section and the shelves devoted to nature and wildlife are on opposite ends of the store. Could composting be the missing link?

While the heap gently heats, I explore the research and literature. The more I think I do for my pile, the more I realize it will do its own thing anyway, if you know when to lend a hand and when to get out of the way.

"Every time you mix pea vines or carrot tops into the soil, you unleash a cascade of biological activity. Insects, mites, snails and earthworms begin tearing the plant material into pieces as they eat their fill, creating residues that smaller organisms can access more easily. Fungal hyphae begin growing through the leaves and stems, excreting enzymes that digest the tough cell walls. Bacteria and other microorganisms colonize the exposed surfaces, absorbing the nutrients that have been released for the plants' growth and activity." – Keith Reid

Another title I've added to my compost bookshelf is *Teaming with Microbes: The Organic Gardener's Guide to the Soil Food Web*. Expert gardeners Jeff Lowenfels and Wayne Lewis have come to see compost in an entirely new light. "While we were out spreading fertilizer and rototilling our garden beds by rote, an ever-growing group of scientists around the world had been making discovery after discovery that put those practices into question. Once you are aware of and appreciate the beautiful synergisms between soil organisms, you will not only become a better gardener but a better steward of the earth."

Describing compost as a whole universe of diverse microorganisms, the authors explain how the benefits derived from adding even a small amount of compost to your garden "are almost incalculable when it comes to managing the soil food webs in your life."

With humus comes all the benefits of the soil food web: decompaction, aeration, better water retention and drainage, and increased retention and availability of nutrients. The authors also contend that after a year of composting and avoiding synthetic fertilizer and pesticides, this new soil life will extend down as deep as eighteen inches. "When you team with microbes, there is no small print to read—and no health problems for you, your family, or your pets. The microorganisms in the compost you apply to your garden, trees, shrubs, and perennials will spread life as far as they can. It is microbial manifest destiny."

Here I thought I was the master of the domain that is my pile. Not true. I am only the minder. As composter and organic farmer Ken Singh puts it, "The microbes in our compost are the best employees I've ever had. They work tirelessly. They don't complain. They never go on strike. By golly, I love 'em! All the networks of fungi and microbes in soil are interconnected. We're part of that, too. One day we'll end up back in the soil ourselves."

The Comfort of Snow

A winter storm churned up the Atlantic coast yesterday afternoon, developing into a nor'easter that left more than a foot of snow in its wake. The morning-after view of the yard from the back porch is positively bucolic. My pile is as pristine as it will ever be, and I leave it to bask in the morning sun, snow crystals glistening across the smooth white covering.

After clearing snow from the porch and shoveling enough of the driveway to back my car out in case I need to, I clomp through the virgin white toward the buried mound. Wind-blown flakes flutter by—with wind chill the temperature is in the low teens. Peering over a deep drift of snow along the windward log wall, I see that the top is pockmarked with vent holes. I gaze through a wrist-thick chimney to find that the heap has subsided, its heat melting the icy snow above to create a luminous cavern; a slick white ceiling looms over the matted surface of wet straw and leaves. It's a cozy scene in an igloo sort of way.

I pick up a few chunks of snow to plug the biggest vents, aiming not to smother my pile but to seal any escape hatches to retain heat, then decide to add more. I want to keep the heap insulated from the ongoing chill and to see just how much more meltwater it can absorb. It's been the better part of a month since I've added water with the now-frozen garden hose. There have been a few cold hard

rains, but this is the first significant snowfall of the season. I know from my deposit of kitchen scraps last weekend and the prodding of the spongy mass with the rebar that my pile is ready to take in a top-dressing of slow-release water.

I slide the wide, flat shovel through some fresh snowpack, then toss the thick sheets of white stuff up and across the pile. It's like heaving a ten-pound wedding cake over my shoulder. A dozen scoops turn the once-rounded heap into a mini Matterhorn of chunky snow. Calculating that each shovelful must hold a cubic foot of snow, which translates to an inch or two of rain, I figure I've just parked fifty or so gallons of water atop my pile. Some will evaporate with the winter sun, but much will melt down into the mix. The snow cover will insulate it, no doubt, enveloping it in a thirty-two-degree thermal blanket, and the slow release of water, drip by icy drip, will allow the brewing compost to continue percolating over the coming days and weeks.

"That is what is so miraculous and so compelling about compost," organic gardener Eliot Coleman writes in *Four-Season Harvest*.

"If you pile up organic waste products they eventually decompose into compost. There is nothing to buy, nothing to be delivered, nothing exotic. No machinery is necessary, and no moving parts need repair. All you need to do is heap up the ingredients and let nature's decomposers do the work. This acknowledged 'best' garden fertilizer is so in harmony with the cyclical systems of the natural world that it is made for free in your backyard." – Eliot Coleman

Some years ago, an Alberta clipper was on the way, with a forecast of below-zero overnight lows through the week. A growing awareness that my pile is its own self-starter and can sustain itself as it consumes itself didn't stop me from embarking on another act of nurturing rather than letting nature take its course. I figured the kind of cold snap that could freeze the local pond thick enough for ice-skating could shut my pile down completely until the spring thaw. So I decided to spread a large tarp across the top, anchoring the corners with four hefty rocks.

The next morning I inspected the pile to find a row of icicles along the tarp's backside. Trapped water vapor had condensed under the insulated covering, then drained along the underside off the lowest point, where most of it froze again.

A gust of wind would occasionally fling one corner of the tarp back upon itself, tossing its rock capstone off the log wall, but over the course of a week the plastic blanket captured the steam and kept the leaves on top of the heap damp, not frozen stiff. A nice feedback loop. The experiment was a success, at least in keeping my pile cooking. Nevertheless, when the cold front eased, I took the wraps off. I missed seeing the heap's mottled autumnal colors and having the chance to check how it subtly shifts and sags. It's the cover-up that always gets you.

On a trip inside to warm up, I empty the fireplace again of its ashes. My fast-filling bucket of kitchen scraps has nowhere to go at the moment except into cold storage in the tool shed, but a blessing of gray ash and charcoal bits could help melt the snow covering and move things along.

From an environmental standpoint, it's hard to justify burning through a cord or so of firewood each winter. Even the best fireplace is relatively inefficient as a heat source, and all that smoke going up the chimney is far from benign.

We all have to make peace with our carbon footprint—and make amends as best we can. The locavore in me likes the idea of using every bit of a tree that has grown in my backyard, from adding its leaves and seeds to my pile, to spreading its chipped-up limbs as mulch, to burning its sawed and split logs in the fireplace. The cost savings, in terms of keeping the thermostat down, appeals to my Yankee thrift. (A cord of oak has the heating equivalent of 159 gallons of fuel oil.) And without a stand-by generator in a region where the power can go out for days, I also appreciate having a source of heat to tap in case things ever really go south.

I like everything about a fire, from constructing it just so to stoking and tending it through the evening. A home fire sustains a gardener's tinkering spirit the winter long. In the words of Charles Dudley Warner, "To poke a wood fire is more solid enjoyment than almost anything else in the world." (A close friend of Mark Twain, Warner also had a way with words; he's the fellow who said, *"Everybody talks about the weather, but nobody does anything about it,"* and *"Politics makes strange bedfellows."* To have listened in on a fireside chat between those two!)

I haven't had to pay for firewood in years—unless you count the cost of having swamp maples taken down by those I hire to scale trees with ropes, belts, and chainsaws. Owning a chainsaw is a manly aspiration I've always found reason not to fulfill. In these

parts, firewood is free for the taking if you know where to source it and have the means to schlep it home. I've helped myself to cast-off chunks of oak left by the utility crews working up and down the street and trundled home loads of fresh-cut logs from neighbors who have had tree work done and have no need for firewood.

Though it took a long while before I could get myself to hand off the big-boy axe to Cole, chopping wood has become a family enterprise. Together, we make a good team, taking turns with the maul, splitting wedge, and axe to turn a whole log into fractionalized pieces with a few, well-measured strokes. This old-fashioned manual labor strikes Cole as thoroughly exotic. It's also a contest between old man and young buck; you can almost see the testosterone coursing up from the steel bit and wood handle into his sinewy young arms. I like being a father who has taught his son how to chop wood. Better yet, he still has all his toes.

The stack of firewood in the backyard is my pile's doppelganger. They're both pit stops for assembled loads of energy-rich carbon. One is on the fast track through the circuit of life and will soon go up in smoke. The compost heap is taking the long way around, a journey with a more lasting reward. Its stores of carbon and other turbocharged amendments are first going to ground, where they will dissipate over time before making their way back skyward, rising toward the sun, whether as a blade of grass or soaring tree that eventually burns bright in a crackling fire.

What goes around, comes around. "Earth to earth, ashes to ashes, and dust to dust in sure and certain hope of resurrection to eternal life," says the Anglican burial prayer.

I bundle up and trudge out across the snowpack. Setting my bag of fireplace ash and furry dander aside, I pick up the rebar rod I keep leaned against the back of the tool shed and pierce the pile a dozen times or so, from all angles and sides, circling its flanks like a caveman finishing off a woolly mammoth. I focus my prodding on places still thick with snow.

I'm wagering that my pile can continue to thrive under its insulated snow blanket, and poking it a few more times will activate any dormant areas. It's a good, quick workout, and before long I've created a score of shafted bore holes for meltwater to soak those dry compartments of leaves that surely surround the small areas into which I've forked supplies of fresh green rotting stuff.

Staying upwind, I flick a portion of the soot across the top of the heap. Chunks of charcoal tumble down the front face, but most of the ash swirls and sticks onto the snow. This albedo of gray will gather sunlight and melt the snow, sinking slowly downward to add carbon and nutrients to the mix. There's a certain symmetry to sprinkling the final remains of the maple tree upon the compost heap that long nurtured it. Just as its leaves have contributed to the soil, so too will the ashes of its limbs and trunk. Besides, I like the look of charcoal gray on white, the poetic juxtaposition of spent fire on frozen water.

I'm generally sparing in the addition of wood ash, a very caustic material, to my pile. "Small amounts are fine," advises Mike McGrath in his *Book of Compost*. "But we are talking small amounts. No more than a cup of ashes mixed into a 4 x 4 x 4-foot bin."

McGrath would rather see wood ash sprinkled across the lawn or garden (instead of lime) to raise the pH level of the soil, which

tends toward the acidic in areas of plentiful rainfall. This I do, walking across the yard to jiggle dustings of wood ash and bits of charcoal from the paper bag. I'm more liberal dusting the lilac and forsythia with ash, as I've heard they like their soil to be on the sweet side of alkalinity.

Quoting Julia Gaskin, a Land Application Specialist for the University of Georgia Extension Service, McGrath further explains that "ash from good quality hardwoods contains a very nice amount of potassium; at least 3 percent by weight. Potash improves root health and strengthens the very cellular structure of plants, helping them resist all kinds of stresses."

Here's more about potash, from Dan Sullivan, soil scientist with the Oregon State University Extension Service: "In the eighteenth century, the benefits of ash-derived potash, or potassium carbonate, became widely recognized. North American trees were felled, burned and the ash was exported to Great Britain as 'potash fever' hit. In 1790, the newly independent United States of America's first patented process was a method for making fertilizer from wood ash (U.S. patent number 1: An improved method of making pot and pearl ash)."

It's fascinating to realize that processing wood ash was once such cutting-edge technology that it will forever be number one on the hit parade of American innovations. Wouldn't you know, though, that we now import fully 96 percent of our potash from abroad, including 1 million tons a year from Russia, at least before sanctions.

I am also intrigued by accounts of how the remains of my fireplace can turn my pile into a rich deposit of "black gold," or what indigenous communities in the Amazon call *terra preta*.

"Researchers say that adding charcoal to soil may provide more benefits for long-term soil quality than compost or manure," I hear from an NPR Science Friday story by Ira Flatow:

"Poor quality soil. It's a problem for farmers around the world. Dirt stripped of nutrients by years of over-farming and chemical fertilizers. Well, there's new evidence that an old farming practice traced back at least 1,500 years to tribes in the Amazon basin can give new life to nutrient-poor dirt. It's called 'black gold agriculture.' The idea is really simple. You add charcoal from burned organic matter to the soil and the dirt holds on to nutrients and produces lots more crops."

Flatow interviews Dr. Mingxin Guo, from the Agriculture and Natural Resources Department at Delaware State University in Dover, who explains: "Charcoal is a fine-grained, porous black carbon, and once applied to soil, the pores allow air to diffuse into the soil. Plant roots need the air to breathe. And in the meanwhile, the tiny pores hold water and nutrients and later supply it to plants. More important, unlike other organic fertilizers, charcoal is very stable and it will not decompose to carbon dioxide. So once applied, it will stay in soil for hundreds to thousands of years."

Intrigued, I read up on biochar and find that it has indeed become "a darling of the climate mitigation movement," writes environmental journalist Kate Wheeling, "a skeleton key for saving Earth."

"Today, biochars are created by burning wood, crop, and other organic waste at high temperatures and low oxygen—a process known as pyrolysis that creates a super stable structure that can lock up carbon," Wheeling explains in *The Daily Beast*. "Of all the 'negative emissions technologies' that the United Nations climate panel said humanity needs to pull carbon from the atmosphere and avoid catastrophic global warming, biochar is currently the only one that is technologically and economically feasible."

The composting industry is already the largest buyer of biochar in the U.S., the article notes, adding, "If the emissions reductions are the front-end benefit for the composting industry, the biochar compost itself is the back-end value."

My pile is a long way from an ancient field in the Amazon jungle, or a mega dairy farm outside of Modesto for that matter, but I subscribe to the theory. As I tend my garden beds and lawn and come across a chunky bit of charcoal coughed up from the compost pile, I think of those indigenous farmers from a millennium ago and thank them for the inspiration.

Mousing Around

It's Presidents' Day. Winter's chill lingers over the landscape. Following successive storms, my pile remains snowbound, like a ship hemmed in by the Arctic ice pack. A spell of record cold has surely driven whatever life still smolders within to retreat into its depths.

I venture out to the woodpile to celebrate the day. From Washington and Lincoln to Reagan and Bush, splitting wood seems a particularly presidential exercise.

"Firewood warms you twice: once when you cut it and again when you burn it." – Henry Ford

While I'm busy with the axe, Miller clambers up a log wall and buries his nose in one of the vent holes that perforate the heap. By the time I turn around to take further notice, he's pawed out a patch of sea grass from under the snow. I don't know what he's sniffing. It could be a remnant of the seashore or aroma of the fermenting kitchen scraps from my last deposit several weeks ago—or he's caught the scent of an interloping scavenger.

There's been no sign of rodents around my pile of late, though it's possible that some furry little critter from nearby has tunneled under the snow. It doesn't surprise me, given the extremes of weather, that a mouse or wood rat would be attracted to the warmth and sustenance of my pile.

It's happened before. A few springs back, Michel shifted around the collection of junk in his backyard (the wood pallets on which he'd stacked all manner of plastic tubs and rusty file cabinets had rotted away), displacing the rodents who'd been nesting below them, and some of the diaspora made their way over to the compost heap.

A round hole in the wood-chip mulch along the bottom of the back fence connected to a same-size bore hole just above the last batch of kitchen scraps I'd tucked inside my pile. Pieces of eggshell and a banana peel lined the opening. Another entryway was neatly scratched out on the ground between two of the wall logs. My pile was turning into a compost condo for rats, with a twenty-four-hour buffet.

I borrowed a slender rectangle of a live trap from the Favreaus and set it alongside the log wall, balancing a cracker topped with peanut butter on the spring hatch in the middle of the metal-wire cage. A couple mornings later, Miller bounded out the back door and nosed straight to the trap, barking with keen interest. Inside, to my surprise, was a small, twittering rodent that looked like a mouse-sized kangaroo. I hoisted the wire cage up out of the dog's reach to take a closer look. The whiskered critter had hobbit feet and a scrunched-over back, like a hamster. It was mousey brown, except for a white ratty tail. It peed through the wire under my scrutiny.

I walked the cage across the yard and over to the car to set it in the back, figuring I'd let the varmint loose when driving my son to school. I don't see pilfering my pile as a capital crime and had no reason to dispatch the animal, at least execution-style. Besides, on the ten-minute drive to his school, there's an empty lot set between the highway and railroad tracks, across a deeply wooded culvert from the marshy inland area of a state beach. The state DOT uses the area to stage trucks and dump spare loads of asphalt and wood chips.

Holding the cage a foot or so off the ground, I opened one hatch and tilted. Out popped the little brown rodent, which sprung across the weedy ground in zigzag leaps and bounds, disappearing lickety-split into some deeper weeds across the way. My son and I were startled by the hip-hop display.

Later, we Googled "Connecticut rodents" and decided our compost lodger was a woodland jumping mouse. Wikipedia tells us it's "a species of jumping mouse found in North America. It can hop surprisingly long distances, given its small size. The mouse is an extraordinary part of the rodent family. Its scientific name in Latin is *Napaeozapus insignis*, meaning glen or wooded dell + big or strong feet + a distinguishing mark. This mammal can jump up to 3 m (9.8 ft) when scared, using its extremely strong feet and long tail."

Back home, I reset the live trap, figuring my chances of capturing a mate were pretty good.

"A casually managed compost pile can become a mouse magnet, welcoming rodents in search of seeds, food scraps, and places for

nesting," I read in a posting on Rodale's *Organic Life*. "Mice living in a compost pile are just doing what comes naturally. Even so, their role in the spread of serious diseases such as hantavirus, salmonellosis, and Lyme disease makes mice undesirable tenants any place where people are at risk of coming in contact with them or their droppings."

The next morning the trap was sprung. Inside was a Norway rat, big and gray and beady-eyed and with no charm whatsoever. I was heading directly to work, so I took the rodent with me, once again with a destination in mind—a small park along the Saugatuck River, a promontory of sorts among the tidal flats. I released the rat in the parking lot, expecting it to scurry into a nearby patch of phragmite reeds. Instead, it high-tailed across the lot and back toward the road. On the other side and up a hill were houses.

Before I'd processed any course of action, a hawk sitting on the branch of an oak tree overhanging the road swooped down and took a stab at the rat as it skittered across the two-lane street. The rodent narrowly escaped, disappearing into the leaf-covered slope on the other side of the road. The hawk took off. So did I. I had no qualms about giving the adorable little jumping mouse a new lease on life, but I still feel bad about unleashing that rat on neighbors.

FEBRUARY

Armchair Composting

A midweek storm has given my son the snow day he's been praying for and me the occasion to make my way out to the pile after shoveling off the porches and driveway once more.

Starting from the back patio, I angle the snow shovel in front of me like the bow of a flat-bottom skiff; it tamps down the fresh snow along the trodden path so my boots don't plunge through to where the snow can reach my socks. This narrow lane soon splits three ways—to the shed, to the stack of firewood along its side, and to my pile. A desire line, landscape architects call these paths of convenience.

Entombed in its sarcophagus of white, my pile is a ghostly outline. I hardly remember what it looks like unbound. The log walls that frame it are stacked with layered flattops of snow, testaments to the successive storms of late. Between them is a saggy crater of white with a stain of brown in the center, a good sign that the composting within has not been brought to a standstill.

I pause to consider the foolishness of what I'm about to do. Weeks ago I was heaping snow on top of the pile. Now I'm going to shovel it off. My muscles already ache from the effort of tending the porches and driveway, but it only takes a moment to slide the shovel across the tops of the logs, flinging the snowcaps to the side.

Groundhog Day has come and gone, again. One of the remarkable things about tending a backyard compost heap is watching how it keeps to its own internal clock and rhythms through each passing season, regardless of current weather conditions. My pile is an endless loop of new life reborn out of decomposition and decay. Like Phil Connors constructively idling away in Punxsutawney, I put a lot of energy into my pile. It's a creative act that offers the prospect of improvement over time yet invariably turns out the same, year after year.

Some years we have more snow or less, some spells of freezing cold or a stretch of mild, almost balmy midwinter warmth that makes it seem as though spring is just a few short weeks away. That's not the case this year, and as a pale, weak sun casts its shadow upon it, my pile is busy steaming away in its cloak of new snow. With this much on the ground, it will be weeks before we see even a patch of grass. Snow, shovel, repeat.

This season's abundance of snow is an anomaly, according to an analysis of regional and global snow decline. In 2000, Connecticut had 61.9 snow days, and in 2022 it had 31.1 snow days. "Over the next 20 years we will probably continue with the decline because there is no indication that our world will not continue to warm," the study's author, Stephen Young of Salem State University, told NBC Boston. He added that southern New England was in the top 10 percent of global snow cover decline over the past twenty-two years.

The winter's snow has kept my composting muscles in shape, at least. Though I don't get in much machine-aided exercise, all the raking, shoveling, and pitching I do around the yard provides

plenty of cross training, especially for my arms, shoulders, and core. Scooping a dense white sheet cake off the driveway or even raking feathery leaves across grass is a workout, one that I enjoy. I read once that music conductors seldom suffer heart attacks—the strenuous waving of hands and arms and batons for hours on end gives them heart muscles of Olympian gusto. Not that I'm calling myself a virtuoso with a rake and a pitchfork, but I can only hope my efforts with my backyard instruments count something for cardiac health.

I shovel drifts from inside the walls and from the front flank, stripping swaths of snow almost to the leaves underneath, then work my way around the heap, carving the white like fat from a ham. Warmer weather is on the way, and this unveiling ensures my pile will be snow-free that much sooner—ready for me to start turning it in the warmth and renewal of spring.

Snow conducting aside, the real reason I'd bothered clearing the path back here was to hang a plastic bag full of kitchen scraps from a spare hook on a ceiling rafter in the shed. The lidded bucket I use to store the waste in-house was beginning to fail the sniff test. So I emptied its vinegary contents into a takeout bag, which will soon freeze, joining two others already in suspended animation.

The one path of convenience I hardly ever bother with is to the tall garbage can I keep tucked behind a large bush along the side of my house. With virtually all my food waste destined for my pile, I fill up the kitchen trash can only every couple of weeks; it's mostly plastic wrappings, paper packaging, and soiled paper towels.

Except for a few weeks after this past Christmas. My parents thought they had found the perfect gift to help spiff up what they consider a backwards lifestyle and sent me a fancy new Keurig machine to replace my sorry old Mr. Coffee. The big box came with a sampler pack of K-Cups.

I admit, the coffee was good and convenient, but it pained me to pluck out each punctured K-Cup and toss it in the garbage. Once, I tried to peel the plastic top off to spoon out the coffee grains like crème brulée. The cup out-engineered me. So I went back to my old setup, which gives me the morning jolt I need and the rich, dark coffee grounds I know my pile savors.

I'm not alone in dissing K-Cups, at least on composting grounds. Though the Keurig really is a better mousetrap, at least in terms of a drug delivery system, inventor John Sylvan told *The Atlantic* in an interview: "I feel bad sometimes that I ever did it. Because the K-Cups are bad for the environment—they are disposable and not recyclable."

To each his own, coffee cup or compost. Every compost heap, by definition, evolves organically, in its own way. My pile suits my backyard and reflects the New England climate and the resources I bring to bear on it, including my own energy and ambitions.

"Building a compost heap can be as effortless or as time-consuming as you want it to be; however you decide to play it, you'll end up with usable compost–the only difference is the time it takes to produce." – Claire Foster

My brother lives in the rural high country of New Mexico. He owns a small ranchette and keeps an old mare in a corral out back, rescued from a shelter, plus a couple goats. His compost heap and its concerns are wholly different from my own. Seeing my pile for the first time, he expressed envy for its fulsome amount of leaves and ample supply of rainwater, as well as the seaweed. He has manure, hay, and kitchen scraps the goats don't eat, but with the arid desert and daytime heat, keeping his backyard heap wet enough is a constant problem, as is keeping the coyotes at bay. Instead of decomposing, his pile desiccates, becoming more a mound of mummified remains than a compost heap. I advised him to consider pit composting, and to locate his buried organics near the water trough for the horse, for easy access to both water and manure.

Closer to home, a nearby friend has a house atop a rocky outcrop, with towering oaks that shade all but a patch of her backyard, on which she tends a small garden of herbs and vegetables. Without the time or inclination to amass a heap of leaves, she instead tucks her garden trimmings and kitchen scraps into a tumbler set up on the side of her house. It looks like a fifty-five-gallon oil drum on a rotisserie and churns out buckets of compost in short order that she spreads across her modest garden.

Just down the road, one of the original farmsteads in town has a barn and open field behind the main house. The owners keep a small menagerie of a few sheep, a couple goats, and a llama in an enclosure near the road. When my son was younger, he delighted in stopping by to reach through the fence and pet the animals, lured by a baby carrot or two. The acreage behind the barn lies fallow,

and a real estate sign indicates that the owners are just waiting for the right price to develop the parcel into new homes. But some agriculture still takes place, perhaps for tax purposes, and a few years ago I was delighted to see the owner spreading truckfuls of leaves collected from the town's fall cleanup across an acre of freshly plowed land, depositing them in long windrows about six feet tall. Over the course of several months, he turned the windrows with a small front-end loader, then spread the cooked-down lot across the field. By the next year the ground had absorbed it all, and it's now a rich meadow of field grass, albeit one with a big For Sale sign on it.

I reference Nicky Scott's *How to Make and Use Compost* a lot, but occasionally the tips skew UK-centric. Particularly intriguing is the chapter, "Choosing the Right Composting System," which leads with a description of the Dalek bin: "The compost bin that most people are familiar with is the plastic 'dalek'-type bin, promoted by local authorities. Millions of these are now in use in the UK. Daleks are lightweight, so you can plonk them down where you want either on earth or hard ground. When they get pretty full, lift the whole bin up—as if making a sand castle—and if you have enough space put the bin down next to your compost castle and fork the top, uncomposted layers back into the bin. The bottom section should be nicely composted and ready to use."

The English love their gardens and long ago raised gardening to an art form. So it should come as no surprise that in the land where your home is your castle, millions of council houses and flats have one of these stubby little castles of compost in the backyard.

Hooch Helps

Despite mounds of evidence in support of composting, both at home and on a community and even industrial level, despite a sweeping cultural shift toward sustainability and, lately, a raft of regulations to spur compliance of this greater good, tending my pile remains a quirky hobby of a habit that brands me as the neighborhood eccentric. As much as my heap stands out in my backyard, it stands for the most part as a solitary, somewhat quixotic enterprise.

For those in the "I want to compost, but..." camp, convenience seems to be the greatest hurdle, and to that I would add cost, comfort, and commitment. Adopting any new habit or practice—on one's own, then culturally—requires a shift in what you're comfortable with, coupled with the decision to stick with it.

Comfort and commitment require a bit of effort when it comes to compost. I admit, to borrow a cliché, there is a fly in the ointment. A few flies, on occasion. Fruit flies, to be specific.

In my experience, the biggest turnoff to composting at home is the prospect of having the kitchen food-waste receptacle become the habitat of fruit flies. Though innocuous and non-biting, these flittering little drones (the *Drosophila melanogaster* of high school

biology texts) can become bothersome if allowed to populate a compost bucket indoors. The solution to eliminating their presence is fairly easy—simply keep the kitchen scraps in a lidded container, and after dumping it out, clean the bucket just as you would a dirty dish, with some soap or a spritz of disinfectant. If the flies persist, perhaps attracted by fruit left out on the counter to ripen, try this, from *Modern Farmer*: Mix a bit of dish soap with old beer, wine, or apple cider vinegar and some water. (They like fermented fruit.) The flies will come for the fruit sugar but get trapped in the soap bubbles.

I'm fortunate to have stumbled across my "hooch" bucket, a tag-sale find that has turned out to be a wonderfully useful addition to both my kitchen and pile. The small plastic container, with lid and handle and shaped like a can, originally served as an ice bucket to advertise a lemony malt beverage. It looks just fine on the kitchen counter and effectively keeps my food waste and its smells in check until I can tote it outside. But if I was starting to compost from scratch now, I'd have full advantage of online shopping, which is awash in solutions for compost storage, including high-tech food-waste receptacles installed as built-ins for new high-end homes.

Cost is another concern, of course, and as a resolute skinflint, I've always rationalized my backyard composting as a way to save money—on garbage pickup, yard maintenance, and gardening supplies. By reducing my kitchen waste stream, I don't see the need to subscribe to the twice-weekly garbage pickup that is the rule of our local haulers. Nor have I ever paid someone to take the leaves that fall across my yard off my hands, much less mow the lawn or weed or sort out my plantings. And by producing so much fresh humus each season, my pile allows me to have a garden, for free,

that needs no store-bought fertilizer and virtually no purchased herbicide or pesticide controls. There is also no need to invest in costly composting equipment, such as a machine you plug in to do what nature does for free.

I'm fully aware that I am an outlier, especially regarding the time and energy I devote to composting—and time is money. There is an opportunity cost to composting, and the hours I spend on it do add up. I sometimes fret that with the time I spend on my pile doing something I enjoy, for free, I could be doing something more remunerative. Like driving an Uber. Still, I figure that the hours I "waste" add up to time I'm not using to spend money on more expensive hobbies. This may not be the case for others, and for those people, other viable options for composting are emerging.

And then there are those who just can't get comfortable with the idea of composting food waste at home. I imagine their biggest concern involves attracting pests, chiefly rodents. Every guide to composting fairly screams with the advice not to add meat scraps, dairy products, or fats to an outdoor heap, for fear of attracting varmints to your backyard. While I admire my pile for its sprawling, unkempt nature and don't blame any critters who want to become acquainted with it, I can understand why another homeowner or backyard gardener mightn't be inclined to tend such a setup.

I'm intrigued by the rise of compost entrepreneurs, often Gen Zers who use bikes to run urban routes, like chimney sweeps of old, the milk vans of a generation ago, or the paperboys of my youth. A service called Curbside Compost picks up food scraps around towns in Fairfield and Westchester County on a weekly basis and

in return, for a monthly subscription fee of about a buck a day, will also deliver finished compost to your door.

I'm glad to have close neighbors who "get" my compost pile. Danute grew up on a small, multi-generational family compound on the outskirts of Budapest, behind the Iron Curtain. Now the mother of four very American girls, she remains a frugal Hungarian *hausfrau* from the Eastern Bloc, and her kitchen scraps are quite unlike mine: hefty stalks stripped bare of their Brussels sprouts, lots of eggshells, potato peels, and hard-pressed pellets of coffee grinds from her French husband's espresso machine. Over dinner, she tells her kids of growing up in a semi-rural household with a large kitchen garden and a variety of fruit trees. Canning the apricots, slaughtering the pig before Christmas, packing the potatoes, turnips, and beets in straw for storage in the cellar. They made the most of what they had and wasted little. Her girls clean their plates at every meal.

Quite different from my own dear mother, who had no further use for even the slightest potato peel—not for soup stock, much less a bucket of festering food scraps to keep in the kitchen. Her favorite kitchen implement? The InSinkErator garbage disposal, a device that scares me to this day. Frozen TV dinners were a staple in our house. Growing up in the 1960s and '70s, we discarded enough pocketed aluminum trays to make a jet airplane.

My mother was raised on a large midwestern farm, where nature had been mastered and awesomely large machines powered through the corn and milo fields. Summer visits to my grandfather's farm were always fun, albeit on an agro-industrial scale, playing on the huge green John Deere tractors and combines and running around the cobalt blue AO Smith silos as tall as Titan rockets.

I don't ever recall seeing any sort of kitchen garden out back, and one day when I brought an ear of corn plucked from the edge of a field behind the barns, my grandfather tossed it aside, saying it was yellow corn fit only for livestock. The vast corn rows along the rural highway that ran by his ranch house were labeled with small metal signs marking their genetic variety and, likely, the type of herbicide sprayed by crop duster or tractor pulling a liquid spreader with a wingspan even wider.

These two worlds intersect in Eleanor Perenyi's *Green Thoughts*, a classic of garden writing. An American who married a Hungarian baron in the decade before World War II, Perenyi writes of "remembering the smoking piles of straw and manure on our Hungarian estate." Urged by her husband to flee Europe in the early days of the war, Perenyi relocated to the Connecticut coast, where she spent the rest of her long and productive life writing and gardening. She had a particular and prescient passion for composting:

> "When I learned about composting after the war, it was a hobby for cranks, and neighbors refused to believe the heaps didn't attract rats. (They don't.) Now that 'organic' has become a catch word, composting has even acquired a kind of mythical status. That is nonsense. It is a practice as old as agriculture, and no civilization has survived for long that hasn't found a way to recycle its vegetable and animal wastes...

Composting was, in fact, general throughout the world until the development of chemical fertilizers, which farmers were brought to believe were all that was necessary to replenish the soil...

You can't buy compost. Neither can a healthy, well-conducted garden do without it. Even if you can't bring yourself to believe in it as fertilizer and use it only in conjunction with chemicals, you still can't do without it, for the very life of the soil itself depends on it. Without the microorganisms at work in compost, soil would literally be dead." – Eleanor Perenyi

Big Ag still rules the range and the supermarket shelf, but things have changed on the home front all across the country, in suburbs and cities alike. Over my lifetime, recycling has had its fits and starts, but increasingly it has become big business, an ingrained personal habit to some, if not a necessity for many others as mandated by municipalities overwhelmed by the trash our lifestyles produce.

The push to reuse, recycle, and eat local is driven by a growing grass-roots awareness of ecological concerns and passions, to be sure, but also by the sheer scale of the food we waste and the cost and logistics of what to do with all the resulting garbage. I read in a recent EPA report that 21 percent of the municipal waste stream in the U.S. is made up of food waste. That's the largest segment of all waste types generated, greater than paper and even plastic. Worse,

the amount of food we throw away has risen nearly 300 percent since the 1960s.

It's easy to track this trend, which started, as many such innovations do, on the West Coast. In 2009, San Francisco became the first U.S. city to make composting food waste mandatory. Much of San Francisco's food waste, an article in *National Geographic* informs, is processed at a compost facility called Jepson Prairie Organics, fifty-five miles east of San Francisco in Vacaville.

"'A lot of wineries in Napa and Sonoma are big buyers of the compost because it has a high nutrient value, so that's a nice way to close out the loop from what we put in our green bins,' said Guillermo Rodriguez, the communications director for San Francisco's Department of the Environment. The compost is also sold to individuals, landscapers, and the highway department. It is approved for use with certified organic soil."

Now all of California is engaged with recycling food waste and organic material. A new state law that took effect at the beginning of 2022 mandates that Californians toss unused food, coffee grounds, and other leftovers into bins they use for other "green" waste, such as garden trimmings, lawn clippings, and leaves. Another part of the law calls on grocery stores, event venues, and restaurants to recover a large portion of their edible food and donate it to those in need. "The goal of the new state law is to reprocess 75 percent of the green waste by 2025," Yahoo tells me. "That means redirecting 17.7 million tons of organic material away from disposal, equivalent to the weight of more than 9.5 million cars."

California isn't alone. New York State also kicked off 2022 by requiring businesses making two tons of food waste per week or more to recycle it. And Vermont already has a universal recycling law banning food scraps from landfills. In the nearby Connecticut town of Meriden, a pilot program involving 1,000 households has them putting their food waste in plastic bags alongside their regular trash. It will be turned into renewable energy and compost, diverting 2.5 tons of food scraps a month from the waste stream. I'm heartened to know that 70 percent of households are participating in the program, and that a quarter of households are separating eight pounds or more of food waste each week.

Next door to me in Norwalk, yet another pilot food-scrap recycling program aims to divert food waste from going to the incinerator, which powers much of the local energy grid. Evidently, the soggy scraps clog up the power plant. A benefit of these more industrial-strength recycling programs is that they accept meat scraps and bones. Even my hometown of Westport has gotten in on the act. The local branch of Sustainable CT has set up a food-waste drop-off bin at the refuse center.

Repurposed from an old wine cooler, my humble hooch bucket seems quite a fitting storage vessel for the kitchen scraps. And though it's not destined for anything so grand as, say, a Napa vineyard, I consider every finished batch of compost I produce from my own backyard a unique vintage in its own right. Each year varies in composition and *terroir*. This year, I suspect, my snow-drenched pile will produce a briny, homebrewed mix of humus, redolent of seaweed and mollusk shell, with a taste of oak tannin, a bit of pumpkin spice, dash of coffee, hint of horseshoe-crab shell, and a subtle afternote of Angora rabbit pellets.

The Big Thaw

The temperature trends upward, producing a weekend of spring-like conditions. After a run of warm, sunny days, the blanket of snow quilted by successive storms is gone from my pile, though the ground around it is still frozen hard. The crown of jumbled salt marsh hay has subsided, and the heap is now a crestfallen soufflé, its center having sunk into itself. At the back, the chicken-wire fence strains to contain stacks of freeze-dried leaves—already two of the galvanized staples that pin it in place have popped off like a fat man's buttons—and a swath of drifted snow turned rotted ice nestles at its feet. Patches of snow hold out in the crevasses of the log walls like retreating glaciers, and the north-facing front remains a crusty mat of frozen leaves.

It dawns on me that the reason the heap has seemed so diminished of late is that I've been standing on a foot of packed snow looking down on it. Subtract that lift and my pile has suddenly increased in volume. Duh!

Today, a Sunday, I will take advantage of the February thaw to plump up my pile. First, I'll give it a good turn, at least the top level, with the pitchfork. Then into the cauldron will go an infusion of scraps and leftovers from the kitchen, including the groaning bags of food waste and rabbit poo from the neighbors that have been hanging frozen in the tool shed. I also have a lidded ashcan crammed with coffee grounds from the closest Starbucks, a plump plastic bag full of shredded office paper, and two bins of salt marsh hay scooped up yesterday from the beach. In all, that's fifty pounds or so of high-octane "greens" and such to stuff into my near-dormant pile, a haul that will surely help nurture it through the waning days of winter.

With so many fresh fixin's on hand, I have plans to dig deeper than usual. Since the front and back sides are frozen thick and the center has caved in, I decide the best approach to the first pre-spring turn is to excavate the core and add enough volume to rebuild the heap's crown. Standing atop the log walls, I use a hay pitchfork to spear clumps of the dankest fulminating sections and pull them out toward the edges, tossing and turning the collection as I go.

I'm following time-honored advice practiced by veteran composters like Jim and Mary Competti, founders of Old World Garden Farms in Ohio:

> "If left to be, a compost pile in the winter will begin to freeze much faster from the outside in. The cold materials at the edge of the pile simply can't get enough heat to compete with the air temperature. But by turning and adding scraps to the middle, you allow the fresh green material a better chance to heat up and help keep the heat distributed for as long as possible. It also adds much

needed oxygen to the core. As a bonus, digging fresh materials in through the winter also keeps your pile from being invaded by animals looking for an easy meal."
– Jim and Mary Competti

As I poke through a tangle of sea grass stems, I spot a fuzzy yellow-green tennis ball—which I take to be the real reason for Miller's nuzzling around my pile of late. I toss the ball across the backyard, and he is delighted to retrieve it, snagging a high bounce off the still-frozen ground.

I dig down, turning up detritus from deposits made months ago, before the snow started falling. Most has already been consumed by the digestive process of my pile. The pitchfork tines hook on a scrap of coffee filter here, poke half an eggshell there. I turn up a small plastic container cup, like the kind you get cream cheese in, tossed in with the food scraps in error. It is already packed with a humus-like loam, which I fleck out with a tap against the log wall.

I burrow as far as my hay pitchfork and straining back will allow, clearing out a hole in the center of my pile big enough to crouch in. At the bottom is a tangled mess of wet, pressed leaves. It's cool to the touch, but I'm pleased to see a clutch of fat red earthworms glued to the underside. Holdouts or pioneers, I don't know, but they will soon be richly rewarded for their perseverance.

I fill the hole first with fluffy gobs of crinkly, julienned office paper, then add half the kitchen slop and rabbit poop—two buckets' worth. I mix a bit with the pitchfork before backfilling with forkfuls of thawing leaf chunks. Over this goes the rest of the shredded paper, then a generous sprinkle of coffee grounds topped with the remaining kitchen scraps. I need to bury this second layer of greens with leaves, so I turn my attention to the bulging chicken wire and compressed stacks of leaf litter straining against it.

The large, crumbly blocks of dried leaves fluff apart with the flick of a pitchfork tine. I toss forkfuls across the top of my pile to restore it to peak form. I get about halfway down the back row of pressed leaves, enough to relieve the backward pressure on the wire fence and to leave an easy-access stash of leaf mold for the next time I need a supply of browns. As a finishing touch, I dump the bins of salt marsh hay across the top, each load discharging with a flourish of sand. Like Beetle Bailey, I've just dug a foxhole only to fill it back up.

A measure of sand is always welcome at my house, whether it's clinging to skin, sandal, or towel or brought back by the bucketful. Regular additions of such freeloaded sand benefit my pile and in turn my lawn and garden. Made up largely of inert particles of rock and other minerals like silica, beach sand helps keep the soil in my yard—which leans toward clay past the root zone—airy and stable. Sand adds heft and no doubt some trace elements and minerals to my pile, and I imagine the granular crystals help grind up the leaves, like sandpaper. Sowing a shovelful across the top of a freshly fluffed gathering of leaves in the fall weighs it down just enough to keep the winds from scattering the leaves back across the yard.

I don't worry too much about overloading my pile and yard with the sea salt that infuses beach sand or seaweed. I'm sure it's a pittance compared with the road salt spread on nearby streets each winter. I know that much from the swath of desiccated lawn along the road that each spring seeks to regrow. Living at the corner of a T intersection means the salt-spreading trucks turn wide in both directions, tumbling a thick layer of rock crystals onto the grass. The use of deicing salts, which also include magnesium chloride and calcium chloride, has tripled over the past forty-five years, leading to a troubling increase in salt concentrations in streams, rivers, lakes—and lawns. Roads in some of the icier regions may be treated with as much as twenty pounds of salt per square yard per year. In truth, a single cup of road salt is enough to treat a twenty-foot-long driveway.

Some years ago I was pleased to make use of a local surplus of construction-grade sand, also free for the taking. In the aftermath of Superstorm Sandy, the town's garbage drop-off and recycling center became the final dumping ground for many of the sandbags used by residents to keep floodwaters from their garages and basements. By the following spring, the mound of burlap bags was still rotting away off to the side. So I loaded a half-dozen of the most intact specimens into the back of my car. I kept one bag to spread on icy patches on the porch and patio steps for the next winter and added the rest, one at a time, to my pile over the course of the growing season. I also spread the empty sacks as a carpet in front of my pile to sop up standing water—the burlap soon melded with the muck of mud season to disappear under my feet. True grit, my pile.

Tools of the Trade

A low-pressure system roared up from the southwest overnight, dousing the region with a cold, drenching rain. Howling winds whipsawed the trees in my backyard, their bare branches backlit by flashes of lightning. February thunderstorms are rare in these parts—or were.

Across town, fallen trees blocked roads, and many neighborhoods lost power. Nowadays, each passing storm extracts its toll on a century's worth of suburban tree growth. Some are demasted, their top-heavy trunks snapped off clean, while others are upended whole, root ball and all.

Seeing a tall tree arching far overhead, the mind wants to picture a matching taproot extending as far and wide underground, but you'd be amazed at how perilous the purchase of an old tree can be—a three-foot-thick trunk supported by the sketchiest of root structures. If a tree falls in the forest, no one hears it, but when it falls across a road and a power line, everybody hears the utility crews when they arrive to saw it up and cart it away.

The trees in my yard are mostly unscathed, suffering only a haphazard pruning that has sheared off any number of small branches and limbs. The trees rely on these windstorms to shed new growth they can't support and to cull old rotting branches they

can, and must, live without. That which doesn't kill you makes you stronger. But there are scars, and recovery, which is my chore for today.

"If you have a garden and a library, you have everything you need," wrote Cicero. Anne Scott-James, author of *The Pleasure Garden*, draws a finer point: "However small your garden, you must provide for two of the serious gardener's necessities, a tool shed and a compost heap." In my backyard, these two necessities are side by side.

The shed, a prefab, eight-by-eight-foot saltbox I had trucked to my property from Amish country several years ago, really is a necessity. My small, cottage-style home lacks a garage, so into the shed go the bikes, lawn mower, and cushions for the patio furniture, along with my collection of gardening tools, the leaf blower and hedge trimmer, cans and jugs of gas and oil, a stepladder, and all the other occasionally useful things you stash in a shed.

I grab the leaf rake, loppers, and small green tarp. I drag the tarp across the frozen yard like a sled, piling it high with fallen branches and twigs, depositing them in a brush pile under the pine tree in the corner of my yard, close by the street. Made up of branches and yard trimmings, twice a year or so the messy sprawl grows to rival my pile in size. Our tall trees shed bark, twigs, and limbs with every storm, and after each I sweep the yard clean.

I've thought of keeping this refuse on site, perhaps by creating a second compost pile, but backyard space is limited, as is my time, and besides, I know I'd be picking out the woody stems all through the next season, cursing as they tangled themselves in

the pitchfork tines. I've considered renting or even purchasing a shredder to chip the sticks and branches, but it's a commitment and expense I've never been able to justify. Besides, it wouldn't fit in the shed. Instead, I whittle away at the brush pile throughout fall and winter by breaking some of the limbs and dead wood over my knee and burning them in our small patio firepit. Cole and I also make a special bonfire in the outdoor firepit by the back patio each January with the incendiary remains of the Christmas tree, feeding in branch by branch for a sparkly show.

The cat particularly loves to hang out by the brush pile, as it's also a favored haunt of the chipmunks that chatter about the yard. But when the stack of branches gets big enough that I worry it will become an issue with the neighbors, I borrow Michel's pull-behind trailer, a ramshackle contraption of plywood and rust, and haul the mess off to the town's yard-refuse center, usually in late spring after I've pruned new from the shrubs and bushes. The occasional nor'easter or other weather calamity sometimes requires a special cleanup.

I also have a large, heavy-gauge tarp, twelve by twelve, brown on one side, gray on the other. I didn't buy it new but came across it while helping my friend Don clean up the backyard of his new rental house down the street. A neighbor two houses over on my previous street, he'd followed the same divorced single dad path as me. We found the tarp under a thick pile of leaf mold in the corner of his yard. Evidently a previous renter had simply dragged it, full of leaves, off to the side and left it there. Seasons' worth of more leaves had covered it until only a grommet in the corner stuck out. After shoveling off the leaf mold, it was hardly worse for wear.

The tarps and their cottony cousin—the tattered bedsheet I use to gather up fall leaves—are part of the arsenal I've assembled over the years to tend my backyard and compost heap. Most of the tools hang on hooks on either side of the double doors of the shed, within easy reach.

A good spade is an essential garden tool, of course, as is a wide-mouth scoop shovel, which I use mostly for moving snow and wood chips. The edges of its aluminum flange are worn razor-thin and peel up at the corners, the result of countless scrapings. Every couple years I pound the curled-up edges flat with a claw hammer.

I'm partial to the old-school metal-tined variety of rake, which is particularly good for teasing out leaves from underneath bushes. I keep a flat-tined bow rake on hand for heavier tasks, including re-arranging the salt marsh hay atop my pile.

I like the fact that in some parts of England a pitchfork is known as a *prong,* and in parts of Ireland, a *sprong.* Mine has five rounded, curving tines and is a real workhorse. Sometimes called a manure fork or a hay pitchfork, it's designed for moving clumpy, bulky stuff like straw, wood chips, or compost and is ideal for grabbing and turning masses of leaves, though not always without struggle.

Because my heap is denser than most haystacks or the bedding of a horse stall, the pitchfork sometimes gets stuck. It's not designed to work in reverse, and tugging it out of a clutch of mashed-up leaves has occasionally caused the pronged metal head to detach from the wood handle. A wooden golf tee hammered into the joint serves as a shim. The strength of that splice pretty much matches the load my own joints can bear in twisting or turning the heap. Better it fail than my back.

I rely on the pitchfork to dig through my pile and distribute gobs of leaves and such until the finished compost sifts through the tines. When I have to trade it for the spade, I know the compost is done.

Some of my tools are store-bought, but the ones I prize most are garage-sale finds, made in sturdier times and well worn. The latter describes my two other pitchforks. Both have four flat tines, one with a long shaft, the other short with a flanged handle. This type is often called a garden fork, and I use them to tease out the most compressed leaves from the sides of my pile, or to twist and turn the tines in a hole, mixing things up. Their straight, sharp tines also come in handy when I aerate the lawn and transplant perennials. I keep both outside the shed within reach of my pile, and as a result the wood handles are deeply weathered, the iron rusted.

Having given my pile a top-level turn a week ago, I have no plans to take a deep dive today. But I can't resist prodding it with the pitchfork. I plunge the bended tines through the crusty outer layer of leaves, teasing the mix up and out of its turpitude. This fluffing up will allow the heap to take shallow gulps of air at least and enable it to soak up more of the spring rains to come.

"Every block of stone has a statue inside it and it is the task of the sculptor to discover it." – Michelangelo

MARCH

Ticked Off About Deer

The best thing about my compost pile? The deer don't touch it.

Resilient, highly adaptable, and just darn Bambi-cute, *Odocoileus virginianus* now makes itself at home in many a suburb, including my own. It wasn't always this way.

"With no natural predators, the deer population has grown from an estimated 12—yes, 12—in all of New England in 1896 (following the years of land clearance for farming) to approximately 100,000 today in Connecticut alone. A recent survey conducted by the Wildlife Division of the Department of Environmental Protection estimated an average 62 deer per square mile in Fairfield County," says local resident Peter Knight.

Knight is a deer expert who awhile back wrote to our local newspaper before a town meeting about what to do about white-tail deer. Sixty-two deer per square mile in my home county is a lot of mouths to feed. As I live in the only one of Connecticut's 169 municipalities that has banned all hunting, the upshot of this perennial debate about the deer "problem" is, invariably, live and let live.

My neighborhood street and its flanking swaths of lushly landscaped yards connect an access road that parallels the marshy drainage ditch along a major highway (I-95) to an upland ridge of granite ledge and hardwood trees that is too steep to build on and, thus, largely undeveloped. Midway between those two refuges of privacy and protection is my property, a convenient way station for deer on their daily commute.

Being a corner lot, it's not practical to fence off, so nightly, sure as sin, I am visited by a stealthy band of hungry deer. I don't know exactly how many and how they're all related, but of late I figure it's a family group of a doe and two yearlings in tow, sometimes joined by a young buck. Because they are furtive crepuscular animals that forage mostly after dusk and before dawn, I hardly ever see the deer in my backyard—but I know them for the damage they cause and the signs they leave behind.

Sets of two-pronged cloven hoof prints are a regular sight across my yard, especially in snow or during mud season. Scat provides more evidence of just how popular my salad bar of a garden is with these rangy ruminants, and of how long and where they linger. Patches of green-brown pellets are scattered like buckshot beside the yew shrubs and evergreens planted alongside the house, in the perennial beds along the side of the yard, or just plopped in the middle of the lawn.

I can't be bothered with deterrents such as wrapping bushes in netting, aside from a favored young tree, or installing strobe lights or sonic devices. I've tried spraying my plantings with a batch of coyote urine, but the nasty stuff seems more effective at driving me from the yard than keeping deer from eating my daisies.

Sometimes the dog and I adjust our schedules and catch a deer or two or three loitering in the pre-dawn gloaming; he'll give chase, their fluffy white tails high in retreat. Usually, Miller stops when he gets to the edge of our property and the deer pause in the yard across the street to stare back at him before loping further away at their leisure.

Mostly, the dog just sniffs and snorts at the deer droppings he comes across in the lawn, which I normally let disintegrate where they fall. Lately, though, I've begun to scoop up the cow-patty-sized dung I come across from the young buck marking my yard as his territory. I've grown tired of him messing up my shoes.

I saw a big one last Halloween night. Joanna from across the street had stopped by to say hello after the kids finished their trick-or-treating. We were chatting on the porch when a huge and handsome eight-point buck appeared behind her, sauntering up the middle of the street between our houses. We figured his nightly routine had been interrupted by the costumed kids and parents parading around at odd hours. He took the festivities in stride and slowly ambled up the road on his way to the woods as if he was auditioning for a TV commercial.

Aside from adding the occasional deer deposit to the heap, I admit to some satisfaction that my pile is the one thing in my yard that I don't have to share with deer. All else seems fair game, including many of the native shrubs I've bought that were advertised as deer resistant. Same with the tiger lilies and tulip bulbs I received as housewarming gifts. The deer pinch off the delicate flowers and leave behind the beheaded stems to remind me who's atop the food chain in my yard. Fortunately, they seem to have no taste for snow-

drops, bluebells, and crocuses, which provide the first and most welcome blooms of the growing season.

Some summers the deer let me grow the black-eyed Susans, phlox, coneflowers, and milkweed that are the staples of my native pollinator garden; some years they don't. The deer pass by the cleome that blossom the summer long and attract so many hummingbirds and bees. They ignore the many cultivars of daffodils, but then again, so do most pollinators.

This litany of woe is familiar to gardeners who share their habitat with deer. Sometimes I think the deer act like a fox in a henhouse and just mow down whatever they can, because they can. I see lots of chewed-off branches—sampled and rejected—strewn across my yard.

Over the years I've added more and more native ferns across the property, and I now have six or seven types at least—cinnamon, hay-scented, lady, royal, bracken, and Christmas. The ostrich is the most majestic, the sensitive fern most likely to spread. I like their primordial look and how they unfurl each spring. I also like how, once the fronds have grown to spread their green plumage like a peacock's, a single frond will suddenly start to sway in the summer breeze, pivoting back and forth to its own syncopated rhythm. I guess I'm a fern guy. Most of all, I admire the defenses they've acquired over the eons. I rather like the fact that one of my ferns, *Osmunda cinnamomea*, has been found fossilized in rocks from seventy million years ago. Nothing messes with a fern, not dinosaurs nor deer.

Our deer problem is pervasive and near epidemic proportions. The other morning while driving my son to school, we passed a large

yard a couple neighborhoods down the road and counted a herd of fourteen grazing on the grass. "Look at that, a new personal best," I told him after slowing to make a quick count. "Add to it," he replied, pointing out five more in the yard on his side of the car. Near as we could tell, they were all does and yearlings.

Aside from having to counter-program my garden according to the whims and wills of this crowd, deer bring with them an even greater concern—Lyme disease.

The symptoms of Lyme disease were identified among a cluster of young patients in Lyme, Old Lyme, and East Haddam in 1975—Connecticut towns just forty miles or so up the shoreline. A year later a state biologist identified the deer tick that carried the disease, and a few years on the popular name for it was coined. Dr. Mark Friedman, a professor at the University of Connecticut's School of Medicine, calls Lyme disease "the great imitator, an insidious infectious disease that is very difficult to diagnose."

In addition to the variety of common symptoms—fevers, aches, and rheumatoid arthritis among them—Friedman blames Lyme ticks for current incidents of disfiguring Bell's palsy and Lyme dementia, which can often be misdiagnosed as Alzheimer's. I've had Lyme disease myself, and Miller came up lame one day and had it worse, though a course of antibiotics got him back up and running before too long.

While deer are not the only carriers of infected ticks, they are essential to the successful reproduction and completion of the lifecycle of over 95 percent of them. They are also by far the largest distributors of ticks—just compare the acreage through which they roam with that of a field mouse.

I've learned to take tick precautions when outdoors; deer have had to adapt to us and we to them. I no longer tromp across an unmowed meadow with the dog, search for errant golf balls in the weeds, or venture off the beaten path at any of the local nature preserves. Even gathering up a pile of leaves in the backyard and hauling a bagful slung over my shoulder to the compost heap is risky behavior in terms of exposing yourself to ticks. It's a pain, but I make a habit of stripping off my work clothes after a session in the yard and tossing them straight into the washing machine. During fall cleanup, a long-sleeved shirt and pants are *de rigueur*—and have the benefit of also protecting against poison ivy, a pestilence I have grown to fear, as with each exposure, the itch grows worse.

The point is, you can't write about a compost pile without taking into account the impact deer have on the garden—and gardener. The best I can say about them is that they were here first, they are handsome, graceful animals, and they have the good sense to leave my pile alone.

Spring Forward

Last night we got a timely nudge toward spring with the changing of clocks. The seasonal adjustment means little to my pile, which keeps to its own time, but the added hour of sunshine after work will give me that much more daylight to spend dithering about the yard.

While my pile's decomposition is largely driven by biological processes that take place within it, especially through the cold, dark days of winter, it is solar power that fuels the final transformation from an assemblage of dead and rotting remains into living new soil. At the moment, the seasonal tilt of the planet's axis relative to the sun is tipping my pile in the right direction, warming it through and through. Soon, the whole heap, not just the top portions into which I've plugged a winter's worth of compostables, will be engaged.

The weather is also cooperating. This past week the daytime temperatures have spiked into the sixties, even seventies—a record for the date. With each successive year, we are rewriting the record books for seasonal warmth. Not only is autumn lingering longer, spring is springing ever earlier. Sprigs of forsythia and crabapple now flower in December, and I have to take care not to trample the frozen sprouts of daffodils that emerge starting in January.

Springtime temperatures have increased in 97 percent of the country since 1970, according to new analysis from Climate Central, with some locations in the Southwest spiking six degrees or more, on average. The early rush to spring also means a jumpstart on mosquito and allergy seasons. Connecticut is already plagued by mosquitoes carrying West Nile Virus, and in 2021, Eastern Equine Encephalitis, a serious but rare virus, was detected in mosquitoes for the first time in the state. I hate skeeters almost as much as they love me.

Allergies I've feared ever since my father started wheezing and sneezing in his forties. Like the need for reading glasses, the onset of allergies often comes with middle age, so each year a springtime sniffle brings a serious case of the dreads. So far, so good for me at least, although I read that allergy seasons are becoming longer and growing more intense. New research from the University of Michigan suggests that by the end of this century, pollen emissions could begin 40 days earlier in the spring than in recent years. Thanks to increasing temperatures caused by manmade climate change, the science also predicts the annual amount of pollen emitted each year could increase up to 200 percent. That is nothing to sneeze at.

Spring has sprung. The crocuses are up across the lawn and garden beds, displaying their cupped flowers of violet, white, and yellow like so many tiny Easter eggs. The daffodil blooms are not far behind, followed closely by the sprightly flowers of the ephemeral, and native, bloodroot. Striving to keep pace are the forsythia, another harbinger of spring, fast forcing themselves to bloom. Two thirty-foot strips of forsythia bushes line the road on

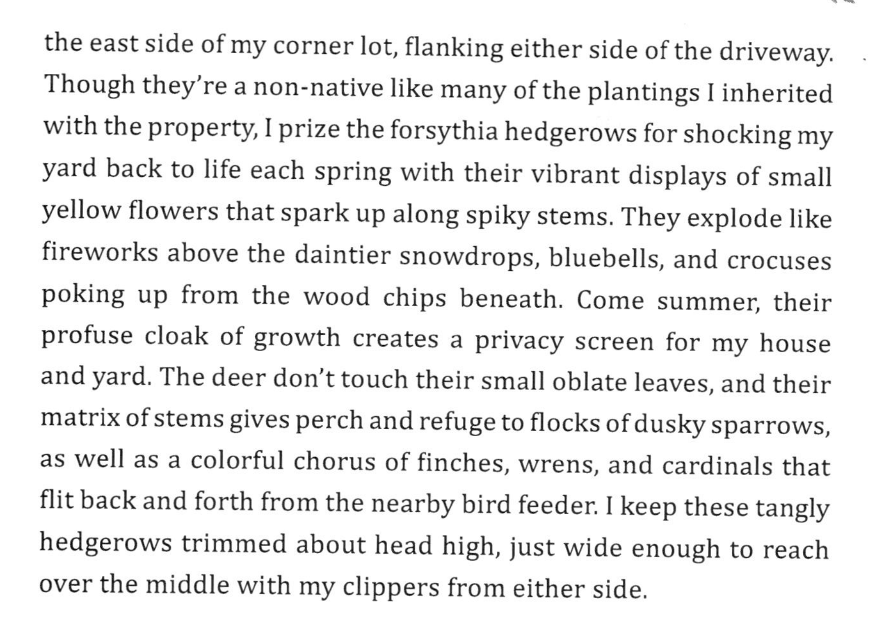

the east side of my corner lot, flanking either side of the driveway. Though they're a non-native like many of the plantings I inherited with the property, I prize the forsythia hedgerows for shocking my yard back to life each spring with their vibrant displays of small yellow flowers that spark up along spiky stems. They explode like fireworks above the daintier snowdrops, bluebells, and crocuses poking up from the wood chips beneath. Come summer, their profuse cloak of growth creates a privacy screen for my house and yard. The deer don't touch their small oblate leaves, and their matrix of stems gives perch and refuge to flocks of dusky sparrows, as well as a colorful chorus of finches, wrens, and cardinals that flit back and forth from the nearby bird feeder. I keep these tangly hedgerows trimmed about head high, just wide enough to reach over the middle with my clippers from either side.

Still, it pains me to be tending so many imports. Over the past few years, I've cut back the forsythia from either side of the driveway, replacing one side with tall Joe Pye weed, button bush, and spice bush and the other with beech saplings, which I'm trimming to serve as a native hedge. Like some oak trees, they keep hold of their dead leaves through winter, making them good privacy screens.

The robins returned this week to stake out their patches of turf, stomping around the lawn and cocking an ear to the ground to suss out juicy earthworms for their coming hatchlings. I read of a 2022 study by Chicago's Field Museum that determined about a third of the bird species nesting in Chicago have moved their egg-laying up by an average of twenty-five days over the past century. The researchers found that the amount of carbon dioxide in the atmosphere over time neatly maps onto larger temperature

trends, which also correlated with the changes in egg-laying dates. "The majority of the birds we looked at eat insects, and insects' seasonal behavior is also affected by climate. The birds have to move their egg-laying dates to adapt," said John Bates, curator of birds at the Field Museum.

The state bird of Connecticut, the American robin, also happens to be the most populous bird in all of America—370 million strong at last count. The robin is already one of the earliest birds to lay eggs; I imagine its chances of surviving the climate crisis are good as long as Americans continue to tend millions of acres of lawns for them, as I am preparing to do today.

Despite the string of warm days, the grass is still dormant; in fact, it's brown and brittle. So why mow now?

The sycamore that lords over the northwest corner of my yard has been shedding spiky seedballs all through winter. Once hard as golf balls, the seedpods that adorn the branches overhead by the thousands are now ripening. Cottony brown fluff drifts across the yard on the gentlest of breezes; firmer winds knock the balls down to the ground, where they disintegrate into dander.

You have to give the tree credit for being so fecund (old-timers know the sycamore as the buttonball tree for its prodigious supply of spawn). Reading up on *Platanus occidentalis*, I find that the seedballs are called achenes, which means "dry, hairy fruit." Each ball contains hundreds of seeds emanating from a round kernel the size of a pea. Each seedhead has a tail, which ripens into silken strands. It's a marvel of design, but on a suburban

lawnscape, the scattered mess is a nuisance. The downspouts of my gutters are filled with fluff, as are the storm drains along the street. Last weekend I fired up the leaf blower to preemptively whisk the pompoms and dander from the gravel driveway; otherwise I'm sure I'd have to spend spring pulling sprouts or, worse, contemplate using weed killer, containing toxins I'd just as soon avoid.

The fluffy seeds are so thick across the lawn that Carl, my across-the-street neighbor, commented on it the other day, suggesting I haul out the mower and scarf it all up with the leaf catcher. Aside from the scandalous unsightliness of so much windblown detritus, I worry whether the covering of seeds will choke off growth of the grass. And, of course, I wonder whether this windfall of organic plant matter, once gathered, could benefit my pile.

I haul the Toro from the saltbox shed and set it in a patch of sunlight to warm the engine block, a trick that seems to help it start better. I last used the mower in late November to mulch the final leaves of fall, and I cross my fingers that it will start up. A check of the gas tank finds it half full, and I worry whether the gas has gone bad. But the engine revs to life after a few tugs on the starter cord, and off I go. Coursing over the brittle brown grass, I fill three tall leaf bags full of fluff and set them against the side of the tool shed. The tricky part will be incorporating this fine mess within the heap in a way that heats the seeds sufficiently to prevent them from germinating. What my pile needs now is a good tossing to air it out and prepare it for the hothouse growth of spring and summer that will allow it to consume itself wholly and fully.

I begin by using the wide bow rake to scrape the salt marsh hay from atop the middle of the heap toward the back, exposing the dank leaf litter underneath. I grab the hay pitchfork to heap forkfuls of musty leaves to the sides and across the front, building up the edges and carving out a trench across the middle nearly two feet deep and just as wide. Aside from a stray eggshell, I see no sign of the kitchen scraps that nurtured my pile through winter, and there is no scent of anaerobic rot from below. Newly exposed, the inner reaches of my pile appear to be just decomposing leaves, warm to the touch.

I scatter most of the first bag of sycamore fluff into the chasm, mixing a bucket of food scraps into the cottony brown fluff, then bury the lot by dragging the salt marsh hay back over the top. I top off the trench with pitchforks full of leaf chunks from the back edge of the heap, the portion regularly soaked with my morning practice of pee-cology. Cleaving these soggy old leaves from the back wall both builds the top of the heap back up to chest height and wipes the slate clean of my daily pit stops.

I have bags more of the sycamore dander and the Tremblays' contributions to add, so I begin another trench-like excavation along the front of the heap. I tease out matted leaf litter from the top to build up a wall higher still, a palisade above the sloping front edge. This is the construction of my pile that I like best, preparing it architecturally for the deconstruction to come. A few cockleshells and a tangled bit of monofilament fishing line are all I unearth as I raise the front and sides to match the new height along the back.

I empty the rest of the first bag of sycamore fluff into the cavity, toss in the Tremblays' food scraps and the green, alfalfa-like hay from their rabbit hutch, then give it all a good mixing with the pitchfork. Turning the bended tines backward, I draw forkfuls of dried leaves from the sloping front of the heap and turn them up and over to fill the trench. Freed from the crush, they expand and unfurl; springs uncoiled. My pile is renewed and recharged. I've dug deeply into it and added a fulsome supply of nutrient-rich organic waste, along with big gulps of air. It still looks like a thick stack of dried, crumpled leaves, but within this cocoon there is a seething riot of new life being created.

Heave and Haw

It's Palm Sunday, and also the official first day of spring. On through the week, the news media has been predicting a nor'easter to arrive by tonight.

The forecasts, as they usually do, started with dire predictions of a foot or more of heavy, wet spring snow, prompting a run on milk, bottled water, and batteries at the local stores. The latest computer models show the weather system staying offshore as it tracks northward up the Gulf Stream, with only a chance of an inch or two of snow starting later today. I tell my son to finish his homework; he hasn't a prayer of a snow day come Monday.

But with the prospect of snow and the distant storm producing high surf on the nearby Sound, I have already laid in my compost provisions, combed from the beach yesterday: a half barrel of seaweed, mashed by the waves and heavy with wet sand. It now sits beside my pile, along with two half-filled buckets of food scraps and a leftover bag of sycamore fluff mulched up from the lawn a week ago. And hanging from a hook in the tool shed is a gift from Don, my down-the-street bachelor friend—a double-wrapped plastic bag of cabbage leaves, potato skins, leek trimmings, and other leftover makings from his St. Patrick's Day celebration. Perhaps this stew will help get my pile's Irish up.

I've also got a vacuum cleaner stuffed with a winter's worth of domestic debris—dog and cat fur, dirt and sand tracked in from outside, dander and dust balls flecked with the down feathers from a pillow fight between Cole and the girls next door.

Most any vacuum-cleaner bag contains a cringeworthy amount of detritus of our own making—sloughed-off skin and hair along with the remains of all the mites and motes that share the interior spaces of our lives. I rather like recycling all that stuff. Dust to dust, as the good book says, with a stop in between to be recycled by my pile.

This morning the craggy brown leaf layer that covers the surface of the heap is damp; smoky wisps of water vapor hint at the cooking underneath, but I wonder if I perhaps overdid it on the sycamore seeds. Sure enough, as I turn out the top edges with the pitchfork, I unearth patches of matted orange-brown fluff, unchanged from the week before, as immutable as Donald Trump's hair.

Fortunately, the sycamore's seedball output spikes only once every few years, and I won't have to deal with the mess again for a while. Shade trees and turf are mortal enemies, and on the playing field that is my pile, this fluff will be no match for all the grass clippings to come.

I am well practiced at borrowing from the sagging center of the heap to build up the edges, then backfilling from the raw front and rear sides. I'm relieved to see the jangled stalks of salt marsh hay, buried just a week ago, are rotting nicely. And once more, the foodstuffs previously tucked inside have done their disappearing act, save for an eggshell or two and the wrinkled skin of an avocado.

The front of the heap is now a raggedy stack of pressed leaves, like shawarma on a spit. I shave off slices of the compressed leaf litter and turn them up and onto the top, once again building the crown up higher than before. Over the past few weeks I've borrowed about three feet from the front scree, and it, like the shelved backside, now forms a near-vertical wall.

Freshly fluffed and refueled, my pile will sit tight for the next week or two. This stretch of spring is a season in waiting, a time to prep the garden beds and plot out new backyard projects. The buds of the trees and flowering bushes are still nascent; squirrels scamper from their nests in the maple trees to sample the budding magenta flowers that tip out the top branches. The lupines are the most recent sprouts in the garden beds; the fiddleheads will be next to unfold. My forty-pound bag of bird seed is empty, the feeder having been overrun by a flock of rapacious grackle. The frugivore cardinals and robins are now picking the last crinkled berries off the privet bushes. Still, there's work to be done ahead of the coming storm, so I lace up my thickest-soled boots and head to the shed for the straight-tined pitchfork, spade, and spare bucket.

It's pothole season, the time of year in these parts when the local road crews switch from spreading salt and sand and scraping snow to plugging the innumerable cracks and gaps and holes that suddenly materialize in the roadway, often just in front of a tire.

The daily cycle of freeze and thaw here in New England conspires to rework the Earth's skin, paved and not. No longer a supplier of prized onions, this land and its silty, sandy mix of sedimentary clay atop glacier-scrubbed bedrock still produces a crop. Out of the subterranean matrix comes an unending supply of what old-timers call the region's most enduring harvest—the Connecticut potato, a catch-all term for the rounded rocks of all sizes that emerge from the subsoil each spring.

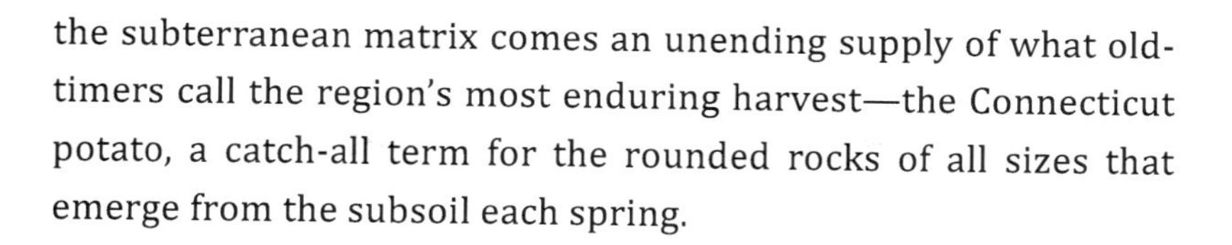

"After having finished grafting roses the gardener finds that he ought again to loosen the baked and compact soil in the beds. This he does about six times a year, and invariably he throws out of the ground an incredible amount of stones and other rubbish. Apparently stones grow from some kind of seed or eggs, or continually rise out of the mysterious interior of the earth; or perhaps the earth is sweating these stones somehow." – Karel Capek

The science tells us that the freezing cold penetrates down into the soil saturated by the soaking fall rains. Stone is a better conductor of heat and cold than the surrounding soil, so the soil under the rock freezes faster than elsewhere. Since water expands about 10 percent when frozen, and the path of least resistance for a rock in soil is up, after many cycles of freeze and thaw, rocks rise up through the mud to the surface. The frost-heave phenomenon helps explain why New England has so many rock walls.

Each spring I get the troublesome stones out of the way by hunting and pecking around the lawn with the flat-tined pitchfork, sharing space with robins doing much the same for worms. As much as

any compost heap, turfgrass needs deep drafts of air and water to thrive, grow thick, and crowd out weeds.

Over the years, I've found that if there's a patch of my lawn that's bare or thinly grassed, chances are that just underneath the surface is a rock preventing the roots from reaching down into the subsoil. As the heat of summer dries the soil, it also bakes the rocks just under the turf, which in turn cook the roots above them. So I step on the pitchfork and drive it into the ground, not only to aerate the lawn but also to use as a divining rod, to hear and feel the clang of metal striking stone. By the sound and vibration of the tines, I can tell what's in the first few inches of the turf, even the size of the rock. As Dr. M. Jill Clapperton said, "When you are standing on the ground, you are really standing on the rooftop of another world."

Most of the rocks, spud-sized, pluck up through the lawn without a fuss, often leaving their indentation intact, which I then fill with a dollop of leaf mold from my pile, packed hard with a stomp of my boot. I stretch the pelt of ripped turf back across the surface, tamp it all down again, and know that I've just added materially to my yard by subtraction and substitution. In place of the dense piece of impermeable stone is a plug of raw organic material, a treat for the earthworms and other hungry creatures that populate and enrich the soil.

The exercise is good for me, and, one plunging, synchronized footstep at a time, I get into the groove of rapid-fire hole punching. I can make twenty or thirty steps in a row before getting winded, or worse, sloppy with fatigue. Some years back, I went on too long and carelessly drove the end of the pitchfork into the toe of my boot,

through the sole, into the ground. Shocked at the misstep, I gingerly pulled the tine back through the leather uppers of my boot, then sat down on a nearby rock to determine the damage. Pain mixed with adrenaline as I plucked off the boot to find a puncture hole in the toe of my sock, already wet with red. I peeled the bloody sock away to find that, miraculously, the pitchfork tine had thrust neatly between my big toe and second, just nicking either side. All I'd suffered was a close call.

I've been much more careful to toe the line with the pitchfork ever since. More and more, the slower I go, stepping on its top edge and driving the row of eight-inch daggers up to the hilt. Deep-tined aeration, greenskeepers call it, and it punches holes down through the impermeable layer of root-stopping hardpan clay that often forms under the topsoil, five or six inches down. As I go, I multiply each footstep by four, the number of tines, and calculate how many individual holes I've made, knowing that each will soon fill with a fresh filtration of organic material, if not from the mulched leaves or spread compost then from the first cutting of grass. I imagine them as pixie sticks for grazing worms and subterranean elevators for fireflies come summer.

The effort may look dorky in a labor-intensive, robotic sort of way, but before long I've aerated a good-sized patch of the yard, usually sticking to the most-trafficked areas. The lowest spots of the yard that once ponded up after a heavy rainstorm no longer do. I credit this to the cracking of the hardpan underneath—and the generous enrichment of humus. Along the way I prod up bucketfuls of loose stones, which I add to the rock wall that borders the northwest corner of my property. The smallest stones I save for

backyard projects like filling in a new posthole. Larger specimens help augment the rock borders that line some of the perennial beds. There is no end of uses for rocks here in Connecticut, nor any shortage of supply.

Sometimes, mud season turns up a bigger surprise. The spring of my second year at the house, while edging the border of a new perennial bed, I came across a jagged tip of granitic rock. I started digging away with enough vigor that the kids playing in the backyard with my young son that day came over to see what all the fuss was about.

There's some Tom Sawyer in us all, and I handed the shovel to the oldest boy of the bunch and invited him to dig in. For the kids, it became a treasure hunt, a backyard mystery, and the chance to show some youthful muscle. Taking a perch on a sizable rock that I had unearthed the year before, I got to opine about how big this new find might be, and where it might have come from—maybe the granite mountains of New Hampshire, carried here by a glacier. Or perhaps from the bedrock outcrop that rises behind the homes across the street, tumbled down this way long ago.

The urge to move rock must be in our blood, a Stone Age impulse. Once the kids had shoveled the dirt from around their treasure, to find it about as big as a beach ball and too heavy to lift, they had to find and use tools—a crowbar from the shed and a couple of long two-by-fours from Michel's inexhaustible home depot of odds and ends—as fulcrums and levers. It was an interesting exercise in backyard-style applied engineering, and somehow, the kids managed to hoist the rock out of the ground and tumble it to the side. It remains in its place years later, the cornerstone of the rock

border of the shade garden next to my pile. It makes a convenient perch from which to ponder the backyard—and remember the day when I actually got a bunch of suburban kids excited about doing manual labor.

Other rocks that have bubbled up to the surface of my yard each spring require stronger backs. A few years later, I had to tap my buddy Don for help, with the promise of a beer or two for the effort, in digging up a large rock that my pitchfork had pinged hard. It lay just beneath one of the barest patches of grass and took a full afternoon to unearth, then roll into place as a sitting stone in the mint garden I keep by the back door. I backfilled the cavity with a whole wheelbarrow full of humus from my pile, and now that part of the yard is one of the thickest patches of grass I have.

Though I try to level out the hollows formed by replacing rock with leaf mold and compost, there's always a certain amount of settling. The turf is a rumpled quilt of dips and swales formed by all these buried pots of compost gold. But my lawn is immeasurably richer for it. It soaks up even the heaviest of rains. And I have more rocks than I know what to do with.

Holy Ground

It's Easter Sunday. Today I will worship not at a church but at the altar that is my pile. You don't have to dig too deeply into the canon of writings about compost to find a spiritual, even mystical appreciation of the process. For some, composting is a near-religious act.

Stu Campbell in *Let it Rot!:* "In the beginning, there was manure."

Steve Jones in *The Darwin Archipelago*: "Adam's name comes from *adama*—the Hebrew word for soil—and Eve from *hava*—living—an early statement of the tie between our existence and that of the ground we stand on (*Homo* and *humus* also share a root)."

The Rodale Guide to Composting: "The compost heap in your garden is an intentional replication of the natural process of birth and death which occurs almost everywhere in nature. Compost is more than a fertilizer, more than a soil conditioner. It is a symbol of continuing life."

Wendell Berry in *The Art of the Common Place*: "The most exemplary nature is that of the topsoil. It is very Christ-like in its passivity and beneficence, and in the penetrating energy that issues out of its peaceableness. It increases by experience, by the passage of seasons over it, growth rising out of it and returning to it, not by

ambition or aggressiveness. It is enriched by all things that die and enter into it."

The high priest of agrarian values and the simple life offers this commonplace about the virtues of composting:

"A person who undertakes to grow a garden at home, by practices that will preserve rather than exploit the economy of the soil, has set his mind decisively against what is wrong with us. He is helping himself in a way that dignifies him and that is rich in meaning and pleasure.

It is apparently impossible to make an adequate description of topsoil in the sort of language that we have come to call 'scientific.' For, although any soil sample can be reduced to its inert quantities, a handful of the real thing has life in it; it is full of living creatures. And if we try to describe the behavior of that life we will see that it is doing something that, if we are not careful, we will call 'unearthly': it is making life out of death. Not so very long ago, had we known about it what we know now, we would probably have called it 'miraculous.'" – Wendell Berry

This Easter I will give to my pile a tithing of fresh green horse manure. Rich in nitrogen and ripe with voracious microscopic decomposers, it will kick-start the heap and supercharge it with phosphorus and potassium, both vital elements to spring growth.

Yesterday, partly to set myself up for a day of gardening chores, I drove Cole's grandmother, my ex's mother, from her assisted-living facility nearby to a horse-rescue farm in the northwest corner of the state. If she had religion, Gigi's patron saint would surely be Saint Francis of Assisi. A lifelong animal-rights supporter, she sponsors a broken-down racehorse now in pastoral retirement. She wanted to see the old filly and hand-deliver a further donation. I was happy to drive her there.

The Vatican is a bit more equivocal on the point person for my pile and me. The patron saint of gardening is Saint Fiacre, but it seems he had an aversion to women, which is why he's also considered the patron saint of those afflicted with venereal disease. Hard to cast yourself in with that lot. I've heard Saint Phocas, or Phocas the Gardener, described as the heavenly protector of compost, because in anticipation of being martyred by Roman soldiers he dug his own grave in his garden so that his remains would be subsumed by the soil. Props to him, but I'll pass—at least for the time being. I keep hearing about the virtues of terramation, the process by which human remains are transmuted into compost, but I'd rather see my life's work as more of a here-and-now enterprise than a "plant-dad-and-see-how-that-turns-out" prospect.

Instead, I made this pilgrimage to a nonprofit manger in upstate Connecticut, a complex of stables and paddocks devoted to giving comfort and shelter to rescued thoroughbreds, retired carriage horses from Manhattan, and the occasional abandoned Shetland pony or miniature donkey. The shelter also gives young girls a chance to groom and ride the horses. Other than that, its chief product is horse poop.

"That's the one thing we have plenty of," said the friendly blue-jeaned blonde who runs the place, directing me to a ten-foot-tall mound of manure in a muddy enclosure behind the barn—a sight for any backyard gardener to behold. It reminded me of Karel Capek's line in *The Gardener's Year*, "Everything that exists is either suitable for the soil or it is not. Only cowardly shame prevents the gardener from going into the street to collect what horses have left behind; but whenever he sees on the roadway a nice heap of dung, he sighs at the waste of God's gifts."

The word manure stems from an Old French root word, *manouvrer*, "to work with the hands or cultivate." Borrowing a thin-tined rake set against the paddock fence, I got to work, forking into my beer-keg tub a rank mixture of horse droppings, rotting straw, and wood shavings. I could only fill the bucket about halfway to the brim before it got too heavy for me to lift, and it's a good thing I remembered to bring along a heavy-duty plastic bag to cover it, or the ride home with my former mother-in-law would have felt much longer.

The manure was a perfect Easter feast for my pile, a hot green counterbalance to the last of the sycamore fluff. I sometimes rue the passage of horses from our daily lives, and I suppose I seek these odd sources of manure as much for the nostalgia they provide as for the nutrition they lend to my mostly modern composting efforts. Clip, clop strikes me as a pace worth following—and cleaning up after. I've read that Westport's onion farmers supplemented the local harvest of seaweed with railroad cars full of manure from New York City. It was a mutually beneficial arrangement. In the 1850s, city leaders commissioned a team of scientists to study what was considered one of the most pressing issues of the day.

Based on population growth trends, scientists calculated that "New Yorkers would be slightly above their nostrils in manure" by 1935, I read on renoarts.com. It's never what you fear the most at the time that turns out to be the real problem. It's a different sort of horsepower that now brings existential dread. I can only pray that my backyard compost pile can play a very modest part of the still-evolving solutions to the climate crisis. Back inside at dusk, I return to the compost scripture and its ancient virtues.

"The ancient Akkadian Empire in the Mesopotamian Valley referred to the use of manure in agriculture on clay tablets 1,000 years before Moses was born," I read on the University of Illinois Extension website. "There is evidence that Romans, Greeks, and the Tribes of Israel knew about compost. The Bible and Talmud both contain numerous references to the use of rotted manure straw, and organic references to compost are contained in tenth and twelfth century Arab writings, in medieval Church texts, and in Renaissance literature."

"The ground's generosity takes in our compost and grows beauty. Try to be more like the ground," said Rumi, the thirteenth-century Persian poet and Sufi mystic.

Speaking of the Divine, there's this from Bette Midler: "My whole life had been spent waiting for an epiphany, a manifestation of God's presence, the kind of transcendent, magical experience that lets you see your place in the big picture. And that is what I had with my first compost heap."

The Rodale Guide to Composting was published in 1979. For gardeners, this seminal work, as thick as the King James Bible, is gospel. Even so, its authors remain confounded by the unknowable essence of their subject, stating, "The entire composting process, awesome in its contributions to all plant and animal life, is probably impossible to contemplate in its full dimensions."

Even pagans—or pagans, especially—genuflect to the glories of humus, or so I gather after coming across "The Gospel of Compost," which won a sermon-writing contest conducted by the Covenant of Unitarian Universalist Pagans.

> "The gospel of compost isn't a story of the permanent triumph of life over death, but of the eternal interconnectedness of life and death, of joy and defeat, of loss and fulfillment. And ultimately, it is a story of love. Love for the world right here, right now, in all its glorious messiness.
>
> Give me your moldy, your stale, your sprouting potatoes. Bring me that wilted, pitiful bag of salad you really meant to eat this time. Bring me your bananas too brown and mushy even to make bread with. Bring me your grass clippings and fallen leaves. Give me the wretched refuse of your teeming refrigerator, yearning to rot free. Give me these, and we will make life itself." – Holly Anne Lux-Sullivan

Amen to that. Maybe I should incorporate my pile as a church, if only for the tax credit.

APRIL

April Fool

Every bloomin' April first, the joke's on me, as I'm reminded of the foolishness that remains my low point as a compost-minded suburban gardener.

A big reason I bought my small home in Westport was the tulip magnolia tree in the corner of the front yard. After noticing a real estate listing in the local paper, I arranged to meet a realtor at the house on the first Sunday in April. I was ready to sign the contract as soon as I pulled into the rutted driveway and saw the beautiful flowering tree in full bloom. Talk about curb appeal!

The house and the rest of the property were a mess. But this specimen of a tree stood out, even though it was besieged by smothering vines and surrounded by spiky barberry bushes and sucker saplings from its own spreading roots. About thirty feet tall and with a canopy almost as wide, it was covered by fluted cups of white flowers tinged with magenta.

Peering through the scrub bushes and stringy saplings that rose from its roots, I could see that the tree's bones were good. Its lowest branches started about waist-high and spread in handy increments nearly horizontally; a perfect tree for my then five-year-old to

climb. Underneath its canopy was a smattering of crocuses poking up through the weeds that spread across what I could tell was once an oval island of tended garden surrounded by grass.

Placed as it was in the front corner of my yard, and that part of the property being on a slight bend in the road, it was the prettiest tree in the whole neighborhood. Approaching my house from either direction, it was as though you were driving straight toward the tree and its blossoming beauty. It was a head-turner, that magnificent magnolia, even if for just a week or so each spring.

After moving in, I pruned the tree of its sucker branches and cleared the ground around it of the barberry and wild wisteria and Chinese bitterroot vines that sought to overtake it. I also dispatched the huge burning bush that crowded it out from the street corner. My son and the neighborhood kids he soon befriended loved to climb the tree's smooth-barked trunk and perch on its low-spreading main branches. For that spring and the next, the tulip magnolia made for great fun and wonderful photo ops, especially in the brief blooming moment, often just at Easter.

For a backyard composter, a tulip magnolia is no great shakes. I raked up the silky fallen petals each spring—they melted into my pile like rice paper—but the seedpods the flowers produced were less welcome, as were the waxy leaves that rained down each fall. Some compost books consider the leaves a nuisance, as they take too long to decompose, but into the mix they went and no harm done.

To restore the garden island the tree grew on and also to give the kids a softer landing in case they were ever to fall from its limbs, I

added a layer of wood-chip mulch around its base, spreading it out to the tree's drip line. I proudly counted how many wheelbarrow loads the ground beneath the tree absorbed, mentally tallying both Safe Daddy points and the kudos for sustainable gardening methods. The tree thrived, as did the kids.

An autumn or two on, a neighbor at the other end of the street took down a towering spruce tree that posed a threat to his house. I drove by just as the tree crew was chipping up the last of the branches and blowing them into a plywood-sided box in the back of the two-ton truck. I told them they could deposit the load of chips in my driveway up the street.

It was like a hundred Christmas trees all ground into a poultice of mulched needles, bark, and sappy chips. The mound was already steaming when I spread load after load of minced spruce across the garden island on which my tulip magnolia ruled. My greatest fear was that I would bury the crocuses and daffodils too deeply and they wouldn't be able to find their way up to the sun come spring.

The following April the magnolia bloomed magnificently. The spring bulbs did well, too. The second spring surprised me—fewer blooms crowned the tulip magnolia. I chalked it up to the vagaries of a tough winter. By the third spring, the tree bloomed only sparsely, and produced small, wilted leaves. Neighbors walking by would stop to chat and offer advice. I watered deeply. That summer, I hammered a dozen tree-fertilizer spikes into the ground all around it.

Then one afternoon, Chris, the fellow who had done all the tree work for me when I moved in, happened by. I'd been impressed by how he handled the massive hulk of the dead old willow in the backyard, saved my roof from the overhanging mulberry trees, and showed no mercy for the swamp maples. Chris stopped his truck in the street, rolled down his window, and, scratching the bandana that always covered his head, gave me the news that he clearly thought I should have known all along: Magnolias don't like their surface roots covered by mulch. The heat from the decomposition cooks them and could kill the tree.

I thought back to a previous fall, sticking my hand in the deep layer of mulch under the tree to feel its warmth. After so many years of neglect, however benign, I thought I was giving the tulip magnolia a warm blanket of freshly made compostable wood chips from which to draw nutrients.

As soon as Chris drove off, I grabbed my wheelbarrow and shovel and removed dozens of barrows full of old mulch from around the tree, spreading it elsewhere in the yard as best I could. I drove more fertilizer spikes into the ground, as penance. But by then it was too late.

In its final spring, the tulip magnolia mustered a few misshapen blooms and a smattering of leaves, most of which shimmered to the ground during a hot spell in July.

I took the tree down that August with a borrowed chainsaw. Its demise, played out over the better part of four years, was slow-motion, everyday proof of my ignorance as a gardener. I'd loved the tree to death, killing it with what I thought was the kindness

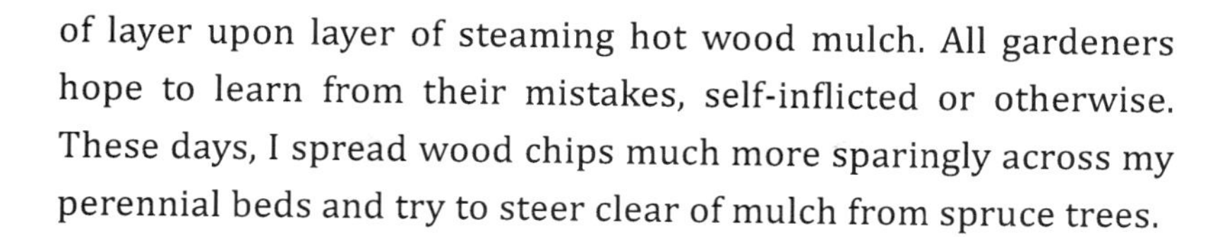

of layer upon layer of steaming hot wood mulch. All gardeners hope to learn from their mistakes, self-inflicted or otherwise. These days, I spread wood chips much more sparingly across my perennial beds and try to steer clear of mulch from spruce trees.

> *"Confess yourself to heaven;*
> *Repent what's past; avoid what is to come;*
> *And do not spread compost on the weeds*
> *To make them ranker."*
> – William Shakespeare, "Hamlet"

I replaced the tulip magnolia with a tulip poplar. I stumbled upon it as a five-inch seedling on the path leading away from the parking lot of a nature preserve donated to the town by Paul Newman and Joanne Woodward, perhaps Westport's most famous residents. The sprout would surely be trampled, so I freed it from the ground with my car key and drove home with its bare roots tucked in the front pocket of my flannel shirt. If I'm to be branded a plant rustler, I can only hope Butch Cassidy would be willing to pardon me for my earnest transgression.

Native to eastern North America, *Liriodendron tulipifera* is a fast-growing, sun-loving tree, but without the weak wood of, say, a willow. I encircled it with deer netting wrapped around bamboo poles for three or so summers, until the top branches cleared browsing height. To keep it from growing through the utility wires

along the main street, I planted it close to the corner of the yard, hedging to the side street and, for company, near a red oak that grew from an acorn.

Five years on, the poplar is now nearly as tall as the tulip magnolia it succeeded and already lords over the red oak. Though it may lack the tulip magnolia's magnificent presence each spring, the poplar does have attractive leaves and, come May, pretty yellow flowers that do, indeed, look just like tulips. The tallest eastern hardwood, poplars can reach 180 feet or more. And while it's no tree for climbing—like redwoods or sequoia, big poplars often don't have limbs until you reach thirty or forty feet up—its straight trunk won't create sightline problems for turning cars. Best of all, it is a hardy living thing that I know will survive whatever foolishness I inflict upon it. Rather like my pile.

Wandering in Place

I celebrate Earth Day by spending a bright, sunny Sunday morning tooling around the backyard. The crabapple and dogwoods have burst forth in full bloom, and the Japanese maples are vibrant with their own budding out, as are the lilac bushes, forsythia hedges, and redbud tree.

I am sure there's a systematic way to add to, aerate, and otherwise mix my pile in the most efficient and productive way possible, a process with inputs and variables that could be modeled by an AI computer program, spit out, and followed—commercial composters take just such a scientific and mechanized approach to their operations. But my pile is artisanal. It's handmade in small batches and sampled throughout the year but mostly harvested in one fell swoop by late summer. The recipe varies from year to year, as does its specific cooking time. Some parts mature early, and most springs I can harvest a wheelbarrow or two of fresh-hot compost for the first sprouts in the vegetable garden or new transplants in the perennial beds or to fill the holes left by rocks I pluck from the lawn through mud season. Creating each new vintage is part art, part science. Mostly it's about mixing air, water, and sundry organic ingredients by turning my pile inside out, in place, with a minimum of fuss and to maximum effect. It's a sport-like hobby, a pastime that engages me both spiritually and physically.

The guidebooks and online sources describe a bewildering array of compost configurations and contraptions, from homemade to high-tech. Structurally, the best description I can find for my setup—a steep, open face of leaves flanked by ascending parallel walls of posted-up logs and backed by a stretch of wire fence—is that it's known as a "log cabin" compost heap. I rather like that. There is a Lincoln Log aspect to my pile, harking back to a baby-boomer childhood spent playing around suburban construction sites and building forts in the woodlots yet to be filled in by new housing.

Life might be easier if I had room for a multi-bin heap or something with removable wooden slats that turns composting into more of an assembly line. But however rustic or homespun my pile may be—customized by site, climate, weather, and owner—I take comfort in realizing it is part of a long tradition, closely aligned with principles first advocated by Sir Albert Howard, a British agronomist and botanist who is considered the founder of the organic farming movement.

According to Claire Foster, "Howard developed what is now known as the Indore process of composting (named after the area in India where he was stationed from 1924), which was based on an ideal of three parts plant matter to one part animal manure. The principles at the root of Howard's thinking are summed up in an unforgettable statement that we could all do well to remember: 'Artificial fertilizers lead to artificial nutrition, artificial animals and finally to artificial men and women.'"

"His principles of layering and aerating are still applicable," Foster relates. "The materials Howard used were animal manures, brush

(twiggy material), straw or hay, leaves and soil, arranged in alternating layers in a wooden bin to a height of 5 ft. A layer of brush came first, followed by 6 in. of plant matter, 2 in. of manure and then a sprinkling of soil. Care was taken to moisten the pile with water while building, and the pile was turned, once after six weeks, and again after twelve weeks. Later, Howard experimented with using human urine mixed with kitchen waste and materials high in carbon such as straw and leaves. The Indore system is labor-intensive, and the heap doesn't reach extremely high temperatures. It does, however, produce good quality compost in a reasonable length of time."

My pile follows the Indore system with a prototypical American twist, which *The Rodale Book of Composting* describes as the University of California method. It's fitting, as my composting has its roots in my days working at the food magazine in California and I happen to be a UC Berkeley alum. "The composting method developed at the University of California in the early 1950s is probably the best known and the most clearly articulated of the rapid-return or quick methods. Turning is essential to the California method, for it provides aeration and prevents the development of anaerobic conditions," advises *Rodale*. The key is to "make sure that the material from the outer layers (top and sides) of the pile ends up in the interior of the new pile."

My favorite part of Rodale's description is this: "For continuously composting household, yard, and garden waste while maintaining optimum pile size, a 'wandering compost pile' is effective. Starting with minimum dimensions of 3 feet high by 3 feet wide by 3 feet deep, this type of heap 'wanders' as fresh ingredients, such as kitchen refuse (minus meat or animal fat), are tossed onto the sloping front face and finished compost is sliced from the back."

In practice, turning my pile reminds me of the old-fashioned flywheels you see pulling taffy in a candy shop on a seaside boardwalk. First I spread the heap open and out, adding air and space, then I fold back in fresh heapings of green and brown from the edges. I think this taffy-pulling bioturbation of my pile is the best way to go about it, as is the log cabin I keep it in.

All this new stuff goes on top
turn it over, turn it over
wait and water down
from the dark bottom
turn it inside out
let it spread through.
Sift down even.
Watch it sprout.
A mind like compost.
– Gary Snyder, "On Top"

At its peak late last year, my pile swelled to a height higher than my head and sprawled across the log walls that sought to contain it. More leaves cascaded over the wire fence along the back and down the front slope onto the lawn. Each time I watered it, or rain or snow fell upon it, the heap shrunk within itself, subsiding

under its own weight and succumbing to unseen forces of gravity and entropic decay. And each time I tucked a fresh batch of kitchen scraps and other organic recyclables into its midst, I gathered more leaves from its messy flanks or the yard beyond to build it up again, as high as my eye. If it didn't always and inexorably settle, my pile would now be twenty feet tall.

Given the heap's sizable volume and stout configuration, it has proven to be a fitting backdrop for all sorts of other backyard pursuits. Like the time Michel came home with a bow-and-arrow set salvaged from the town recycling center. As I recall, it was some years into the Tremblay girls' *Hunger Games* phase. I set up a target in front of the heap for the girls and Cole to practice, figuring my pile was as hard to miss as the side of a barn. It was, until a wayward arrow sailed past the heap and over the stockade fence behind it into the neighboring yard. The quiver was returned to the dump at Danute's request.

Cole and I put the heap to better use during his brief career as a Little League pitcher. Almost exactly matching the dimensions of a batter's box and plenty high, my pile made a perfect backstop. As it did during Cole's inevitable boyhood infatuation with playing soldier. While we largely managed to escape the paintball craze, my son and his chums all had airsoft guns, toy pistols that pump out plastic BBs with just enough velocity to plink an empty soda can off its perch in front of the heap. By the time Cole was done with the gun business, he was a fair, and safe, shot—and I was picking those darn neon pellets from the compost and yard for years. As much as anything else, my pile is made of memories.

But that was then. Time for this old wandering heap to get a move on.

Over the past few months I've narrowed the pile's footprint by nearly half, pulling a wide swath of leaves that once bulged against the back wire fence up onto the top and cleaving three feet or more of compressed leaf litter from the once-sloping front. True, I've prodded and poked and probed the heap with the iron rebar rod, perforating it to allow air and water to penetrate its inner recesses. But up to this moment, I have only stirred the top half of my pile and nibbled at its edges. I have yet to get to the bottom of it, where fresh air and water are needed most to spur on the decomposition process.

So here I stand, pitchfork in hand, ready to set my pile on the next leg of the journey it started last fall. The Big Dig. After setting out the day's additions—a week's worth of kitchen scraps, a fresh bucket of seaweed and salt marsh hay, and the last scraps of sycamore fluff—I stick the pitchfork, face down, into the base of the near vertical front edge to pull a wedge of moist, matted leaves from underneath. It's the first light of day these leaves have seen since autumn. Each forkful I heap onto the back of the pile teems with squiggly worms and glistens with flecks of mica from the beach—any trace of the seaweed they traveled in on is long gone.

Before long I've undercut a strip across the front edge to create an overhang of pressed leaves and tattered seagrass, which I pluck off with the upturned tines of the hay pitchfork and add to the top of my pile as high as it will stand. After shaving this scraggly brow, I have a new wall of old leaves, which I undermine once more. I

pull a line of moist leaf litter from the bottom across bare ground toward my feet to form a berm, about shin-high along the front, a step clear of the heap. I scrape into the new trench behind it some dried leaves from the corners and creases of my pile, then add the last of the sycamore fluff and the fresh kitchen scraps, mixing them with a few twists of the pitchfork.

As I work to fill the gap, the newly created overhang of pressed leaves quivers, then tumbles down to bury the trench with an avalanche of crumbly brown detritus. It's pleasing to see so much mass move on its own. Gravity is your friend when moving heavy loads. You want to move the most stuff with the least energy the shortest distance where it can spend the longest amount of time. This applies to the churn and turn of my pile and to its dispersion later this summer. Also, pretty much everything else in life.

Better yet, newly exposed is a rich, dark, moist cache of proto-compost, the result of a winter's worth of deposits that have nourished the heart of the heap. I tease the mix with the pitchfork and consider harvesting a few shovelfuls to spread across the vegetable garden, having read recently that tomato plants thrive under such unfinished, nitrogen-rich compost. But after some further digging, the batch seems too raw, and besides, I'm still nearly a month away from the last frost and planting time.

I pull more of the steamy scree toward me with the pitchfork, forming what amounts to a terminal moraine at my feet and creating a new trench just behind it. Leaning further in, I unearth a pocket of pristine leaf litter. Bone dry and compressed into a tightly wadded stack, the leaves lie beneath the digging I did into

the center of my pile throughout winter. This time capsule has repelled any intrusion of water or rot and now resists even the tines of the pitchfork. It must be from a load of whole sycamore leaves, which have since flattened out. I tease the leaves apart and spread them across the bottom of the newly created second trench, then cover them with loose stems of salt marsh hay. Released from their suspended animation, they will now fully join the fungible, seething mess.

I keep digging until the inner wall of the heap tumbles down once more, sealing off any further progress toward the core. I add some kitchen scraps and rabbit-hutch leavings to the mix and pull more leaf mold from the crest to cover them.

Backfilling this way shrinks the top of the heap enough to prompt me to walk around the rear side. Reaching over the chicken-wire fence, I fork out a half row of crusty leaves to restore the heap to shoulder height. Facing south and exposed to the warming sun, the rich vein of humus-like compost along the backside is thick with earthworms and now within easy reach.

I finish the hour's work by tidying up the front of my pile with a rake, restoring its facade to the slumping heap of leaves it always appears to be. The log walls make good markers, and by that measure I reckon I've tossed and turned nearly the entire front half of my pile from top to bottom. A good portion of the hard-pressed lower reaches has now become the fluffed-up top.

When time comes to mow, I'll repeat this inside-out, upside-down process, working the back of my pile to add grass clippings; then

I'll dig inward from each flank. It will take me another month or two to get to the very core—to the small mound of compost reserved from last season that served as the activator—but from today to harvest, my pile will be a most active heap, powered by high-octane green. Keeping it aerated now becomes my chore.

April showers are on the way, and the refurbished heap is ready to soak up all the rain it receives and settle back into itself. By taking two steps forward and a half-step back, it is newly suffused with air and freshly mixed organic material—primed for productive decay. And so my pile and I wander in place through the seasons.

Turf Wars

The transition from winter to summer, while always inevitable, has so far been tentative here in coastal southern Connecticut. I woke this cold Saturday to the news saying, "The National Weather Service in New York has issued a freeze watch for Westport and area which is in effect from late tonight through Sunday morning."

A few cases of seedling frostbite aside, this week in late April marks the tipping point when the season begins to shift, delightfully, from cold and dormant to warm and growing. Two days of showers have doused the landscape, and my pile. The forsythia are ablaze in yellow, the fiddleheads and other ferns are spooling upward, and the most precocious of the perennials are emerging from the deep wood-chip mulch I worked so hard to spread last fall. The purplish stalks of the peonies and coiled thrusts of the hostas and lilies of the valley are all sprouting too.

In the vegetable garden, rhubarb leads the charge, unfurling elephant-ear leaves from its robust ruby-red stems that thrust up through a mound of last year's compost. A friend who was raised on a New Hampshire farm once told me that rhubarb grows best on a pile of compost, and I've long followed that bit of farm wisdom. Other self-seeding garden habitués, chiefly the cilantro and arugula, are also sprouting like weeds.

The trees that arch over my yard have yet to issue leaves, though the wine-dark flowers of the maples dapple the grass, driveway, and car. As I was visiting the backside of the heap this morning, slight movement on an old leaf caught my eye. I poked before realizing it was a slimy slug, about a quarter-inch long, affixed to one of the crinkly maple flowers that have rained down upon my pile from the overhanging swamp maple. Scanning the top of the heap, damp with morning dew, I could see that virtually every fragment of maple flower had its own slug curled upon it. I can only guess the slugs were lured out of hiding by the flowers' sweet nectar. Who doesn't like maple syrup?

Soon the lawn will be littered with countless winged maple seeds, helicoptered down from the branches above. In the midst of all this nascent green growth, my pile seems a doughty old relic, though from the early-morning puffs of steam vapor I see rising from its mounded top, I know that it is churning and burning within. Between a busy week at work and spring-cleaning chores at home, it's been a struggle of late to devote attention to yard work. But that's the nice thing about tending a backyard compost heap—there's not much of a deadline involved, and my pile makes few demands upon me, other than those I put upon it.

I hadn't planned to spend much time with my pile this weekend, but that changed yesterday evening when Carl stopped by with a gift from his wife, who teaches at a local preschool. Sarah had guided her students in a special project the past few weeks: making compost in plastic tubs. The kids combined kitchen scraps they brought from home with what I expect was the giggly fun of a class trip outdoors for dirt and leaves and any earthworms they could

find. The assignment now completed, she'd brought the grade-A bins of compost home to pass along to me. My pile continues to benefit from the largesse of others.

Carl also mowed this morning and deposited the clippings at the base of my pile. Three weeks ago he spread a forty-pound bag of fertilizer across his lawn and watered it in with his sprinkler system. I have yet to cut my grass and he's already mowed twice. The first cutting he mulched back into the lawn, but his turbocharged turf has since grown so fast that he had to mow again, this time with the grass catcher to placate Sarah and her white carpet.

It's true—the grass *is* always greener on the other side of the street.

I have some qualms about adding Carl's chemical-laden contributions to my compost but decide to welcome the fresh green material to what is largely still a pile of old brown leaves. I figure the synthetic cocktail that makes up store-bought fertilizer will, in time, break down into its elemental parts and ultimately be absorbed by my pile to a more natural end.

Chemical fertilizers for lawns have only been around since just after World War II. Tom Andersen documents the devastating impact excess nitrogen and other manmade pollutants have had on the water quality (and life) of Long Island Sound in his book, *This Fine Piece of Water*: "In the Northeast United States, each acre of fertilized lawn is covered with an average of 134 pounds of nitrogen a year... Nitrogen that occurs naturally in the soil is taken up by plants for growth only as it is needed, but chemical nitrogen

dissolves easily in water, and anything not used immediately by the grass is washed away in the first rainstorm." Less than half of all the chemical fertilizer applied each year is actually absorbed by plants.

Nitrogen is the most abundant element in the air we breathe, and, as a solid nutrient, it promotes foliage and overall growth and is what gives grass its dark green color. Phosphorus spurs root development, and an adequate supply helps grasses develop drought tolerance. As with nitrogen, an excess of phosphorus in runoff leads to algae blooms, which can prove toxic to both aquatic ecosystems and humans. The third main component in synthetic fertilizer, potassium, promotes disease resistance and aids in the production of flowers and fruit. The rest of the ingredients in a typical bag of store-bought N-P-K fertilizer are fillers to keep the granules from clumping and a bunch of trace minerals, among them magnesium, calcium, sulfur, iron, manganese, zinc, boron, and molybdenum. All are helpful in the proper proportion, though I'd need a textbook to help explain why and to what extent.

Of course, these minerals and nutrients occur naturally, and assembling them in organic abundance and dispatching them throughout my backyard is a big reason why I tend a compost heap in the first place. My pile is nothing if not magnanimous, and mostly welcoming of begged, borrowed, or gifted supplies. It's also a very capable buffer, a backyard biomass factory that I trust to process all manner of raw ingredients into a finished product that is, if not wholly organic, then at least ultimately safe and homemade. So into the mix Carl's hopped-up grass will go. Having tousled the front by stepping it forward while mining the

seam of rich pressed leaves at the back, the crown of the heap has sagged into itself. I'll work the top today; any old additions of food scraps or shredded paper or seaweed are buried deep within.

While I strive to be the best organic backyard gardener I can be, I admit to occasionally falling back on more drastic, manmade solutions when circumstances seem dire enough. A few years ago, my lawn was under mortal attack. After grassing it with the initial renovation of the property, the newly expanded greenscape grew well for several seasons. Then whole patches of turf started to scrape off with the gentlest of raking, or even the scuff of a shoe or paw. Before long, I could roll up the dried thatch like a throw rug. Every time I dug underneath the sod I would find the cause: the subsoil was chock-a-block with short, fat, dirty-white grubs, the larval stage of the scarab beetle. The grubs chomp through grass roots like so many micro-mowers, scalping the lawn from below. At its worst, my lawn attracted foraging skunks, who dug into the turf to feast on the grubs, inflicting further damage.

I considered applying a combination of milky spore, a bacterium, and nematodes, an organic option that relies on a beneficial parasitic creature, to do the dirty work of killing grubs. But these are expensive solutions and require just the right conditions and timing. After much agonizing, I bought a bag of commercial grub killer from the hardware store and spread it across the worst parts of the lawn. I kept dog and son and all others at bay for the better part of a week. I'm sure there was some collateral damage; earthworms seemed in short supply for a time, and it may have been my imagination, but the fireflies that rise from the turf each July also seemed muted that year.

I have since realized that the pesticide was likely a neonicotinoid. The most widely used insecticides in the U.S., neonics have been shown to play a major role in population-level declines of bees, birds, butterflies, and freshwater invertebrates. Mammals, too. Such insecticides are systemic, meaning they are absorbed by plants, making the entire plant toxic. Neonics can linger in soil for years, causing long-term harm.

Though neonics were banned in the European Union years ago, the Environmental Protection Agency didn't release its final biological evaluations until 2022, "confirming that three widely used neonicotinoid insecticides likely harm roughly three-fourths of all endangered plants and animals, including all 39 species of amphibians protected under the Endangered Species Act. The EPA's assessments marked the first time the agency has completed biological evaluations of any neonicotinoids' harms to the nation's most imperiled plants and animals," reports the Center for Biological Diversity. Synthetic fertilizers, which I'm sure were used by the landscapers I'd hired to grub out the superficial tree roots and rototill the yard for seeding, can effectively sterilize the soil, which throws the natural balance of bugs out of whack. It can take up to five years after going organic to get the soil biology where it needs to be, and I'd lost patience. I wish I knew then what I know now.

"Trust the natural to perform its own insect control and watch the songbirds and the tree frogs and the box turtles and the friendly garter snakes return to their homes among us." – Margaret Renkl

Last weekend Carl sprayed his gravel driveway with Roundup and had enough left in the canister to offer a spritz or two for the weeds coming up through my own gravel driveway. Relying on chemicals to treat and care for our yards is made easy by the longstanding habits and practices of our culture, but I waved him off. When my fingertips grow raw from plucking weeds from the trap rock, I prefer to use a homemade concoction of high-strength cleaning vinegar, salt, and dish soap to douse and desiccate the interlopers.

Though I admit to being a killer of weeds, often with extreme prejudice, I prefer it to be a fair fight, relying on hand-to-hand combat rather than chemical warfare. Over the years I've spent countless hours hand-weeding my garden and lawn as I amble about the yard throwing tennis balls for the dog. I'm on the lookout for the "tall poppies," as the Aussies call those that stand out from a crowd. My springtime eradication effort focuses first on dandelions, their seeds having blown into my yard from the untended slope up the street. With a dandelion digger in hand, I've become fairly adept at gazing about the ground in a sweeping, non-focused manner, like a beachcomber, stooping between tosses to stab the fork-toothed prong into the soil to flick up as much of the taproot as I can get before Miller returns with his slobbery ball.

Hand-weeding is a meditation on foot, a touchpoint that appeals to both the hunter and gatherer in me. The dandelion digger is elegantly simple—it was my first tool and perhaps that of humankind, too—a sharp stick with which to pry a plant and root from the ground. One of my earliest memories is of heading out to the backyard at my childhood home in Lincoln, Nebraska with a paper grocery bag and digger. My mother promised me the lordly

amount of one dollar to fill up the bag with dandelions—come to think of it, it was my first paying job.

An import from Europe that has since become ubiquitous on American soil, dandelions have been gathered for food since prehistory. Every part of the plant is edible; I've tried adding the jagged leaves to salads but much prefer the peppery taste of my feral arugula to the bitter dandelion. The bending and squatting is also a fair amount of exercise, and while my goal is to search and destroy rather than forage for dinner or dollar bill, over time my perambulations have allowed me to avoid using chemical herbicides. Then there's this:

> "What better way to get over a black mood than an hour of furious weeding." – May Sarton

I try to let the pollinators sup on these early blooms, but hand-weeding dandelions and such before they go to seed short-circuits more widespread infestations, and their fresh leaves and cloddy taproots make especially nutritious additions to my pile. That said, if I could train the deer to like dandelion salads, I'd be all set. (The one dandelion I could come to like? Researchers in Ohio are growing a dandelion from Kazakhstan with a particularly long taproot. It can be harvested every six months to extract latex from its milky sap. Goodyear sees it as a potential replacement for tropical rubber trees, which take seven years to produce the latex needed to make rubber for tires.)

It's not so much that I favor a monoculture—my motley lawn is a meadow-like mix of annual and perennial rye, fescues, bluegrass, bent grass, poa annua, and other types of weedy "grass." Clover is welcome, to feed both bees and the turf, which it benefits by adding nitrogen through the symbiotic hookups of its roots and bacteria in the soil.

I do battle with other weeds, among them creeping Charlie, clover-like oxalis, wild strawberry, violets, speedwell, certain sedges, pungent field onions, and garlic mustard. It's mostly a stand-off, to keep them from infiltrating the flower beds. Wild strawberries top the list of pollinator-friendly native groundcovers, with violets not far behind. They co-exist with the grass, dotting the lawn with their myriad flowers of yellow and blue. Another favorite that I mow around is blue-eyed grass; I've even begun to transplant clumps of this compact perennial that crop up in the yard to create clusters among the flower beds. It's also a native, and I like its delicate blue-violet flowers. I'm less forgiving of an obnoxious newcomer, hairy bittercress, an early bloomer with spikes of spring-loaded seedpods. Most pernicious of all is that old lawn villain, crabgrass. Derived from millet, the first grain cultivated by man, crabgrass was brought to America by immigrants from Eastern Europe as a crop. It's certainly come a cropper since. The sneaky weed grows low to the ground, often just under the whirring blades of a mower, and produces copious amounts of seeds that, if left unchecked, remain active in the soil for years, just waiting for the right conditions to germinate.

Going to Weed

The thing about keeping a fertile, organic yard for native flowers is that opportunistic weeds take full advantage as well. Each season seems to bring a new, virulent arriviste onto the scene. Any bit of disturbed ground, like a patch of the perennial bed where the wood-chip mulch has worn thin or the bare soil of the vegetable garden I harvest each fall, can provide them an opening. Just the other day I searched online to ID a new mystery weed and came across a *Better Homes & Gardens* listicle about "33 Lawn & Garden Weeds." I have plucked twenty-nine of them from my yard. Do I get a prize?

Last season, hidden behind the big-leafed rhubarb on its mound of compost in the back corner of the kitchen garden, a two-foot-high spike of a weed released its cottony seed puffs to the wind. I'd missed spotting the plant—I'm pretty sure it's American burnweed—because of its similarity to the perennial sunflower I'd planted in the vegetable patch for the bumblebees. I was alerted to this impostor's presence, too late, by the fluffy white seeds I found enmeshed in the fine-gauge plastic bird netting I'd strung above the chicken-wire fence to deter deer.

Sure enough, I've been plucking its descendants from inside the garden and along its outer borders all spring. They're easily riven

from the loamy ground, but all it would take is one survivor to extend the battle to next year.

I'm also on the lookout for remaining traces of the lesser celandine I rooted out late last fall. I've since learned that it was brought stateside as an ornamental. On urbanecologycenter.org, forester Caitlin Reinartz tells what happened next: "In Cleveland, Ohio, lesser celandine was planted in flower beds of (just) two residences in the 1970s. It escaped the confines of those two yards, and less than 40 years later, it had taken over nearly 300 acres of parkland along the Rocky River, with 183 of those acres having lesser celandine cover more than 50 percent of the ground, leaving little room for native vegetation."

One self-invited "weed" I've taken a liking to is rose campion. I noticed it creeping up the street a few years ago and more so when it established itself in the roadside strip along the Grissoms' house, below the rough rock wall that contains their well-manicured lawn. It sends up slender stalks of cerise flowers that bloom from June into August, their vibrant hue reminding me of the bougainvillea of Southern California. Late in the summer Carl asked for help weeding the untamed patch; his back was acting up and he wanted help in sorting out what to keep and what to pull, to make enough space for cars to park with at least one wheel off the street. I was happy to clear the common weeds, all going to seed, from the tiger lilies and irises growing along the base of the rock wall. I left the recent arrivals in place but snagged a few sprigs of the rose campion growing closest to the pavement, figuring they'd be roadkill anyway. I wanted to try them out in the driest, sunniest patch of my perennial bed, near a yucca plant I'd been gifted by the Favreaus.

The pretty plant prospered in that spot, needing no help from me, and I've since let its offspring spread along both roadscapes of my yard and even take root in the beach-stone pathway that leads from the driveway to the front door. It's also known as bloody William, but because its gray felted leaves always came up as lamb's ear on my plant ID app, I never knew exactly what to call it until recently, when I saw "*Coronaria*—rose campion" for sale at a local nursery—for $24.50 a container. I now like it even better. As I like to tell Cole (with apologies to George Washington Carver): "A flower is just a weed growing in the right place."

A most unwelcome new scourge is Japanese knotweed, which I've spotted in the garden bed along the side street that leads to the access road for I-95, some 800 yards distant as the crow, or seed, flies. Between the state road and the freeway it parallels is a long line of this pernicious invasive, taking over the no-man's-land between the two roads. The state DOT crews whack at the weed by extending a scything blade over the guardrail. They spray it too, likely with an industrial grade herbicide. Both treatments are to no avail—other than leaving an unsightly trail of brown and withered, but not dead, broken stalks and a stain of glyphosate that no doubt drains through the concrete storm sewers to discharge directly into the Sound.

Once this import takes root, it will not easily, if ever, give up its claim to conquered ground. Leave a single rhizome in the soil and the knotweed will return, hardier than before. A gardener knows how contagions begin and how difficult they can be to root out.

MAY

As the Worm Turns

It's May Day, and my pile is now fully six months old. As much as the first half of its life was dominated by dead brown leaves, the race to its fruition as finished humus will now be driven by mass infusions of hot green grass clippings.

Today, a Sunday, is the day I mow the lawn and add its surplus trimmings of minced grass to the heap. True, over the past few weeks I've stuffed it with Carl's artificially enriched clippings from his juiced front yard. But those contributions are just appetizers for the main course. Today marks a tectonic shift for my pile, in shape and composition.

A backyard is by its nature a passive, back-of-mind kind of place. But sometimes, like this week, which is unfolding with the full bloom of spring, it demands attention. After days of drizzly rain, the sun now fills a cloudless sky, kick-starting the greenest of growth across the landscape. The leaves on the maples have burst out, the garden ferns unfolded, and the grass is thickening and surging upward. They say you can almost see some types of grasses, like bamboo or switchgrass, grow in real time. Right now, that seems true of my lawn.

The dog's delighted when I trundle the lawn mower out of the shed. His trick is to deposit the tennis ball just outside the path of the mower, entreating me to retrieve it and toss it yonder for him to fetch. It's a sport for both of us; I try to bend down to grab the ball without pausing or veering the mower offline. When Miller drops the ball he waits for my reaction—if it's in the path of the coming lawn mower, a hand gesture from me is all he needs to dart in and grab the ball to reset without slowing me down. In all the years of us playing this game, I've only mowed over a couple tennis balls.

Aside from learning how to throw a ball, ride a bike, swim, and perhaps read, I've been mowing as long as any other thing I've mastered. I jumped at the chance to show my dad I was big enough to take over mowing duties for him as a kid and, as I grew older, made spending money by taking care of neighbors' lawns each summer. The occasional mom-sponsored dandelion hunt notwithstanding, mowing was my first job, and it probably will be my last.

Mowing is a simple, rewarding task that I enjoy. It's a fair way to get some sun and some exercise, especially if you never bother to fix the belt that once self-propelled the rear wheels. Though it is now just dead weight, I wheel my trusty red Toro around the yard like a matador, raising the front wheels just so to skim an exposed root or pass over and along a rock border to protect the whirring blade without having to edge with the hand shears or electric trimmer.

Walking behind a mower leads one down interesting paths of thought. Just ask Forrest Gump. It's a rolling Buddhist prayer

wheel of a meditative act, a squared-off labyrinth that leads to a vanishing point as the final strip of ankle-high grass gives way to a uniform plane of green etched by the tracks of the mower wheels. I always feel better about myself after I've finished mowing the lawn.

Many gardeners fret about giving over precious backyard space to turf—and I agree—but what lawn I keep is as well-trod as center field at Fenway Park—by the dog and me playing our constant game of fetch, and before that by tossing a baseball, football, or Frisbee to Cole. Close-cropped grass makes all the difference; the ball bounces high and true, our footfalls are firm, and no doubt the robins have better luck procuring worms for their chicks.

The only exceptions are a set of patches in the middle of the lawn I leave to thrive as micro-meadows. Two of these are thick with clover, for the bees; the other is a particularly fetching swath of fescue fast going to seed, as a visual treat for me. As I let most of the plantings in my yard come to fruition, it seems only fair to allow this stretch of turf to do the same.

Last summer I didn't mow this patch of lawn until September, allowing it to turn all the way to hay. Now I see sprigs of black-eyed Susan and blooms of small white wildflowers cropping up among the grass, which appears to be the delightfully named "spring beauty." Even better, I find that the low-growing perennial has tiny underground tubers that can be eaten just like potatoes, which is why another name for it is the fairy spud.

In 2020, Cambridge University stopped mowing a section of a King's College lawn for the first time since being laid down in 1772.

Three seasons on, the area was transformed into a true wildflower meadow supporting three times more species of bugs and birds and bats than the remaining lawn. Interestingly, eighty-four plant species were counted during a sampling, though only thirty-three of them had been sown. Nature always finds a way.

Changing how and what we mow can result in as much as a tenfold increase in the amount of nectar available to bees and other pollinators. That's what the backers of a fast-rising initiative called No Mow May are finding. Launched in 2019 by the British conservation group Plantlife, No Mow May asks gardeners to leave the lawn mower in the shed and let your lawn grow to help pollinators at a crucial time of year. Appleton, Wisconsin was one of the first communities in the U.S. to adopt No Mow May, with 435 homes taking part starting in 2020. Researchers studying the impact of this citizen-science project found that No Mow May lawns had five times the number of bees and three times the bee species of mown parks. To keep local officials and neighborhood biddies off your back, Bee City USA suggests making these pollinator patches look intentional rather than neglectful by mowing a border between the uncut grass and natural plantings. Other studies have shown that mowing every other week, or better yet, every three weeks, can also boost bee abundance. I've pledged to myself, and the bees, that I'll just give the lawn a trim, then let it grow for the rest of the month.

Even with the newly sharpened blade set on high, the Toro's catcher fills quickly. I mulch most of the clippings back into the lawn, as backyard botanists advise, yet have to stop a half-dozen times when the mower chokes to detach the hopper and dump the moist,

fragrant clippings at the base of my pile. As usual, I set out my other contributions—a week's worth of kitchen scraps from next door and of my own, as well as two small plastic bags of shredded paper from the office. The crinkled white strips—processed from wood pulp and with a carbon to nitrogen ratio halfway between a maple leaf and sawdust—make a fine counterbalance to the chopped blades of lush green grass.

However much I labor, my puny efforts pale in comparison to the true workhorses of my pile: earthworms. They are peerless in chewing their way through rotted organic matter and turning it into humus. Truth be told, new soil is mostly worm poop.

The worm castings that dot my still-dormant lawn in March and April are ample evidence of this handiwork. Gutwork might be a better term to describe the inexorably dogged pursuits of the class act that is *Oligochaeta*.

"A worm is an animated intestine," Steve Jones writes in *The Darwin Archipelago*. "The body is hollow and filled with fluid, with a long digestive tube down the center. Aristotle described worms as the 'Earth's entrails.' Cleopatra decreed them to be sacred animals and established a cadre of priests devoted to their well-being. [As Darwin's book put it], 'All the vegetable mould over the whole country has passed many times through, and will again pass many times through, the intestinal canal of worms.'"

Worms are the invertebrate backbone of my pile, and I strive to make the heap hospitable to the herd that resides within it. Lowly as they are, worms are near the top of the compost bio pyramid, and they lead the way in deconstruction, perforating the layers

with their burrowing and churning out countless castings along the way.

In form, my pile may look like a heap of leaves mixed with green trimmings, but it functions more like a coral reef, with worms standing (or squirming) in for the multitude of polyps that crank out limestone deposits that, in turn, become the base of an entire ecosystem. Worms do the same on dry land, and what they leave behind is soil.

Charles Darwin may be legend for loftier theories now, but during his life, he had a lot to say about earthworms. His book *The Formation of Vegetable Mould Through the Action of Worms, With Observations on Their Habits*, published in 1881, sold even better than *On the Origin of Species* during Darwin's lifetime. His thesis? "It may be doubted whether there are any other animals which have played so important a part in the history of the world as have these lowly, organized creatures."

Jones explains the math. "An acre of rich and cultivated ground is riddled by five million burrows. Half the air beneath the surface enters through burrows, and water flows through disturbed soil ten times faster than in unperforated. In an English apple orchard they eat almost every leaf that falls—two tons in every hectare each year. In the same area of pasture, they can munch through an annual thirty tons of cow dung."

The earthworm is an eating machine that is a marvel of simplicity. It uses sand in its gizzard to grind up tiny bits of organic matter and the bacteria, fungi, and nematodes that live on it. Moistened with a spit of liquid calcium carbonate, this microscopic gruel travels to the worm's intestine, where bacteria digest it, sending nutrients into the bloodstream and everything else out the back end.

Vermicastings are 50 percent higher in organic matter than soil that has not traveled through an earthworm. "This is an astonishing increase and radically changes the composition of the soil," write soil food web experts Jeff Lowenfels and Wayne Lewis. A worm's ability to increase the availability of nutrients "is about as close to alchemy as it gets."

> "The worm's digestive enzymes unlock many of the chemical bonds that otherwise tie up nutrients and prevent their being available. Thus, vermicastings are as much as seven times richer in phosphate than soil that has not had been through an earthworm. They have ten times the available potash; five times the nitrogen; three times the useable magnesium and they are one and a half times higher in calcium. They shred debris so other organisms can more readily digest them. They increase the porosity, water-holding capacity, fertility, and organic matter of soils. They break up hard soils, create root paths and help bind soil particles together. They cycle nutrients and microbes to new locations as they work their way through soil in search of food. A noticeable worm population is a clear sign of a healthy food web community." – Jeff Lowenfels and Wayne Lewis

Even so, it boggles my mind to know that earthworms are accidental tourists in my pile, having landed on these shores from the baggage of European settlers and the ballast of the ships that brought them here. The glaciers of the last ice age, a mile thick in these parts 10,000 years ago, scrubbed virtually every soil-dwelling worm from much of North America. To the Old World earthworm, this truly was the promised land, and in they rushed, colonizing new ground at 30 feet a year. Some soil scientists lament how dramatically they've displaced native creepy crawlers, both above and below ground, as they munch through the forest floor. That may be true, but they'll always have a home in my backyard compost heap. As E. O. Smith says of earthworms, ants, and their ilk, "Each species is a masterpiece, a creation assembled with extreme care and genius. They are the little things that run the world."

After starting his epic study of worms at his estate in the English countryside, Darwin went off on the Beagle and came home with *On the Origin of Species*. I will stick closer to home and tend to my pile and the worms within it, as they do the real heavy lifting in turning a heap of leaves and recycled greens into rich new earth.

Trunk Full of Junk

It's a fine time of year to be a gardener in the southwestern corner of Connecticut along the shore of Long Island Sound. The landscape is alive with green growth. Oaks, maples, and other hardwoods are beginning to leaf out, casting pollen far and wide, coating cars and nostrils alike. Along the roadsides, pink and white dogwoods bloom in enough profusion to prompt an annual festival.

I tend to the yard on a cool, damp Saturday morning, nipping and tucking the native sand cherry and spicebush, spot-weeding the perennial garden beds and lawn. With the grass bright green but slow to grow, I'm not overly tasked and can stroll about the yard, plotting.

Tomorrow is Mother's Day, traditionally the date by which it's safe in these parts to plant annuals without fear of frost. I've already seeded several rows of lettuce and kale in the raised garden, which are sprouting nicely. Early this morning I went to the annual plant sale organized by the Westport Garden Club to bring home flats of tomatoes and basil.

To prepare the small, fenced-in enclosure of my kitchen garden for planting, I pluck weeds and thin the self-seeded herbs (the cilantro is profligate this spring), filling a small plastic tarp that I drag over to my pile. Across the street, the Favreaus weed the tidy flower bed beside their front door. Each spring, they plant trays and trays of begonias, petunias, alyssum, and other annuals, but must first clear the springtime weeds and grasses that perennially jump the Belgian block border of their lawn. They've filled two blue recycling bins with ripped-up weeds, and I take the dirty mess off their hands. The bins are heavy with the culled clumps of fast-growing, opportunistic plants, most still clutching a dense filigree of dirt in their veiny roots.

It can be a gamble adding spring weeds to a compost heap. They produce massive amounts of seed in short order, which keep developing even after the plant is uprooted. I'm confident my pile will cook the nascent seeds to sterility and consume any undesirable weedlings, so the Favreaus' gift of heavyweight organic green matter is a welcome addition. Weeds are loaded with nutrients snatched from the soil, and to recycle their ill-gotten gains through my pile and return them to the garden as fresh, weed-defeating humus makes me feel I am truly sowing what I reap.

"Compost wants to happen. The more eclectic the list of ingredients, the better the compost. That is only logical. The plant wastes that go into your compost heap were once plants that grew because they were able to incorporate the nutrients they needed. So don't pass up any weeds, shrub trimmings, cow pies, or odd leaves you can find. If you mix together a broad range of plants with different mineral makeups, the resulting compost will cover the nutrient spectrum." – Eliot Coleman

While at this point in the season I've worked the top and front of the heap, I have yet to fully dig into the backside, aside from leaning over the wire fence with the short pitchfork to strip-mine a layer of damp, matted leaves down to the ground. Now the heap is free and clear of the fence and I have room to help my wandering pile move forward by taking a step backward.

After prying out the remaining staples and peeling away the wire fence backing, I begin as usual by tugging apart the crown of the heap, shifting old rotting material up from the middle reaches. It's no coincidence that my pile is precisely as wide as I can reach into the middle with a pitchfork from either side and from back to front. I pull out forkfuls of rank, rotting organic matter and moldering leaves, stacking them along the front and sides. I may have been too cavalier about adding Carl's grass clippings on steroids to the top of my pile, or at least in laying it on so thickly. The smell is just short of an anaerobic stink bomb, and I'm glad to be giving my pile a good airing out.

This freshly dug cavity across the top needs a good supply of raw brown matter to offset the existing rot. I cleave tightly bound clumps of old leaves from the bottom edge of the back of the heap. Each new piece of the ragged wall seems to come with a spring release, the sheaves of leaves expanding to reveal a marbling of decayed salt grass stems and shredded paper, some pieces still so pristine I can read numbers off the crinkly slivers deposited last fall.

These clutches of fusty browns help me fill the yawn while steadily pecking away at the back wall. I wonder how far I can dig into its foot before it collapses. Most of the fun in a game of Jenga is just before the structure tumbles down.

I step away from the backside and plan my way into the cliff-face like a rock climber plotting his ascent. I undercut the bottom further, pulling out compressed tufts of dried leaves first laid down last autumn, to form a berm that stretches a step back along the entire backside. I keep plucking away, the heap stout enough to resist my attempts to undermine it until I've carved out a cavity a foot or so underneath. I fill the chasm with alternating layers of young weeds and old leaves, brown with green, then drag tumbled pitchforks full of more rotted remains down from above.

Spotting an eggshell fragment tumbling down from the newly exposed back wall reminds me to collect the Tremblays' kitchen scraps in their ashcan and my own half-full bucket, and I add their contents behind a deeply edged steppe about halfway up the shaggy backside.

As I'm digging a wide hole along the steaming back-center of my pile to bury, in layers, the rest of the messy weeds and kitchen waste, another neighbor from down the street comes by with his dog, a female pit bull that is vigorously friendly with my own mutt. I can never seem to recall the fellow's name, but Miller and I are well acquainted with sweetly snarling Abby.

My property, being a double lot on a corner, stands out among the postage-stamp yards of most of my neighbors. Having a sociable dog who's often out and about with me makes our yard a popular way station, part dog park, part playground. "Everyone around here calls your yard 'Miller U,'" a neighborhood wag once told me, her young and very big, very frisky Siberian Husky in tow. Helping socialize her pup wore even Miller out. When they were younger,

he and Abby would chase each other, running in circles around the house. Their all-time best was seven laps, if you count the two when Abby caught up to Miller and he in turn gave pursuit. They are no longer as frisky, but there are still track marks at the corners of the house, including a shortcut through the pachysandra.

The backyard trampoline is also open to the neighborhood kids, and it's not uncommon for me to come home from work to find a mom or nanny sitting on the picnic table next to the trampoline watching their kids jump and tumble around, getting tired for bedtime. Fortunately, over the years there's been only one broken wrist among the young bouncy set, and, to my relief, no lawsuits. Any squabbles among the dogs usually involve a tussle over a tennis ball and are soon forgotten.

As Abby the pit bull sports with Miller, my neighbor watches me work my pile. He says he has his own pile of leaves raked up into the corner of his backyard and adds that he kept a bucket half-buried in the ground nearby into which he put some kitchen scraps. But he's never combined the two elements and so is interested in my efforts.

Set alongside the tool shed and trampoline, my pile has grown in size and stature over the years and is now a fairly prominent feature of the backyard. I like showing it off, though having someone look on as I disgorge buckets of kitchen scraps and pitchforks of weeds into the heap makes me a bit self-conscious, like having an audience watch you clean the fridge. My pile has long been more of a Private Idaho than public performance art. But I enjoy delivering what amounts to a compost tutorial, a podcast for one.

I had hoped to dig deeply enough into the rear of the heap to reach a pocket of finished compost for the vegetable garden. But unlike the more easily and often worked front side, I am stymied by what seems to be an impermeable wall of raw brown organics not yet ready for distribution. It's not often I give up on my pile, but it seems to be too much work to get to the good stuff today.

Whether to humor me or bide his time until Abby is tuckered out, the neighbor stays through to see me top off the heap with leaf litter from the front side, covering the rest of the spring weeds and restoring the yin-yang balance of old browns and new greens. Will he return home and commit to turning his own pile of leaves into a full compost heap? Or will he tell his wife over dinner about the eccentric down the street who keeps his rotting garbage in a backyard dump?

On Sunday morning, after a phone call to dear old Mom, I visit the backside of my pile and discover that the rear slope has calved off in a cascade to reveal a cleft of garden-ready compost. I can only surmise that yesterday I'd stopped just short of this pocket. Overnight, my pile has exhaled and coughed up just what I need, in just about the exact proportion I need it, for my rows of newly planted vegetables and herbs. Thank you, Mother Nature.

After scaring away a robin and two grackles that have come to feast on the unearthed deposit, I examine the scree more closely. I spot a shard of avocado husk, still clinging to its stick-on label, and a withered mango pit and toss them aside. A fragment of horseshoe crab shell goes back into the mix, along with the dozens of red worms that cling to the loamy earth tucked inside its carapace.

These stouter chunks return to the pile to cook, but the avalanche has let loose enough thick, crumbly compost for me to scrape up with the pitchfork and fill the wheelbarrow.

The compacted, dark earth looks like the bricks of peat I've seen being cleaved from Scottish bogs. Tossing the clumps into and against the metal walls of the barrow is enough to fracture them into bits with the texture of spent coffee grounds. I finish by scooping up looser chunks with the garden spade, and trundle the heavy load of moist, dark organic matter—new soil—across the yard and through the wire gate fence to the vegetable garden.

Garden writer Vita Sackville-West mused about the ideal recipe for top-dressing and potting soil. "Goose guano and the soil thrown up by moles both had their advocates" in the eighteenth and nineteenth centuries. "Today, the John Innes compost is recommended: two parts sterilized and sifted loam, three parts peat, two parts sand, to which you may add an ounce of hoof and horn per bushel, and some crushed charcoal."

I shovel my own special brand of compost, sans hoof and horn powder, in and among the newly planted tomatoes and basil and along the rows of sprouting lettuce, arugula, and kale. I'm also happy to leave any store-bought peat moss to thrive in some distant bog, where it can perform its time-honored task of soaking up and storing away more carbon than even a tropical rain forest. I lift up the elephantine leaves of the rhubarb to spoon more compost underneath, close to the stalks, which look just like ruby-red celery, and try not to bury the strawberry vines running rampant with ripening fruit.

I spoon more compost across the bare ground, skirting some fledgling milkweed and the self-seeded cosmos. Danute brings by a bag of potatoes from her larder with pale sprouts emerging from the eye buds. I plant a few cut-up wedges in the loamiest part of my garden and plan to add the rest of the bag of wrinkled, sprouting spuds to my pile. She goes home with some rhubarb stalks and a handful of strawberries to make her favorite pie.

Returning the empty wheelbarrow to the compost heap, I decide to really dig into the backside. This is where my sweat equity pays off. Like a miner working a seam of coal, I turn out clump after clump of peat-like proto-compost, diverting the crumbliest forkfuls to the wheelbarrow for the garden and casting the coarser snatches atop the pile to be broken down further under the summer sun.

Having seen my handiwork when she came by with the potatoes, Danute returns with two large pots she'd like filled so she can start some basil of her own. The crumbly loam I'm excavating isn't quite humus yet, nor a total substitute for potting soil, but it will make a good amendment and filler for the bottom of her vats. Consider it the first pour, a sample of all to come.

I wrap up the day's work by unspooling the hose to douse my garden. The tender plants could use a drink, and watering in the compost allows it and those living things it contains to meld with the soil it's landed upon. Over the years, wheelbarrows full of compost have added many inches of new earth to these raised beds, so much so that I've had to raise the pavers and untreated lengths of lumber that give it structure. The little fenced-in plot was

once an unused and unloved patch of weeds—the ugly backside of the house. It is now a highly productive nook that grows dense with fruits, berries, and greens for much of the summer.

The backside of my pile, a stratified stack that I've been chiseling away at for weeks, has over the weekend been replaced by a terraced mix of old organics and lusty new contributions, a hanging garden of decay with plenty of breathing room in which to rot away. I clean up around the edges, front and rear, grooming my pile back into a rough-hewn pyramid shape with steeply sloping faces of crumbly rotting leaves, flecked with grass clippings turning from green to yellow. The front presents an attractive facade as handsome as any compost heap could be. And I've stuffed the trunk full of junk.

Aerial Assault

It's a breezy Saturday, the start of Memorial Day weekend. I wander out to check up on my pile, first scaring off the resident robin couple that tromps across the top in search of worms and other snackables, then disrupting a squadron of hover flies that wafts above it in the warming, sun-shafted air.

Like a monadnock of the desert or jungle, my pile is its own stratified ecosystem. Its ragged summit and steep, flanking slopes are rife with life. I step behind the chest-high log that stands at the rear corner to pee away some coffee, skirting a colony of mushrooms that has sprouted at its base. Midway up the heap, a solitary soldier ant marches across the jumbled scree; a loose leaf jostles, and a foraging red worm wriggles into view. A lanky green spider has woven an elaborate web, pinning its strands to the top of the wire and corner log. I catch a midge-like fly in front of my face with a clap of hands and flick the gnat from my palm onto the silvery matrix. The web quivers, and just like that the gnat is in the spider's embrace, being spun into a bundle of silken thread.

High on the craggy heap, amid damp hummocks of sycamore seed fluff sprouting green like chia pets, more mushrooms rise from the leaf litter. These ephemeral growths are, in fact, flowers. Mushrooms sprout to release packets of fungi genes that seek to

alight on a new source of food to conquer and consume. These numbers are staggering. Each mushroom issues forth millions of spores. So many are airborne at any given moment that I likely take in ten spores with each breath, or so I read. I imagine it's many times that amount when I'm face to face with my pile.

Every time I stick a pitchfork or shovel into the heap, I'm entering another world, one that is dominated by fungi. Like plants and animals, fungi—along with molds and yeasts—constitute a Kingdom of their own. Together, they are the most common organism in the world.

"Without fungi, all ecosystems would fail," writes mushroom impresario Paul Stamets in *Mycelium Running*. The terrestrial world would be overwhelmed by dead plants and animals, as fungi, along with the even more unfathomable bacteria, are largely responsible for breaking down organic matter and releasing carbon, oxygen, nitrogen, and phosphorus into the soil and atmosphere, *Brittanica* informs me. In fact, various species of these decomposers have already started their work in the lower-hanging branches of trees before the leaves even hit the ground.

Hidden within my pile is a vast spread of fungi, each type working their own specialized lanes to create an overlapping network of connective tissue known as mycelium. Stamets calls it the "wood wide web" and says that in a patch of forest floor the breadth of my pile, there are more root endings in the mycelium than neural synapses firing in my brain. He'll also tell you that as much as 90 percent of land plants are in a mutually beneficial relationship with mycelial networks, which help them absorb water and nutrients

as well as build immunity from disease. Not only that, but climate researchers say the vast network of fungi beneath our feet stores over 13 gigatons of carbon around the world, equivalent to more than one third of yearly global fossil fuel emissions.

My compost heap is a node in this global network. Do I worry about being turned into some kind of fungal zombie by breathing in my pile? Not today, for I have other backyard chores to tend to as the holiday, and summer, get underway.

I may be operating under the influence anyway. James McSweeney, author of *Community-Scale Composting Systems*, points out that a species of soil bacterium (*Mycobacterium vaccae*) found in compost releases serotonin, which is thought to relieve depression, anxiety, and mania, among other modern maladies.

Having flowered with countless clutches of winglet seedlings called samaras, the female maple trees in my yard let fly their progeny virtually all at once. One day my driveway is clear, the next it's a fluttering carpet of seedpods. So thick do they fall, I need to turn the wipers on to clear the windshield before backing out of the driveway.

This windfall is a marvel of nature. It's a spectacle to see the winged seeds helicopter through the air, a lesson in evolution and aerodynamics that always delighted my young son, especially when I clambered up to the rooftop to toss whole clouds from the gutter along the front porch.

So many seeds rain down across the property that I haul out the leaf blower to breeze them into piles—this is one use of the noisy

device that makes sense, poetic justice even. I take satisfaction in turning the motorized fan of the spewing little two-stroke against the wind-driven diaspora, using their own propulsion to blow them into easily collectable piles.

Here in the Northeast, maple trees are early winners in the ecological battleground created by manmade climate change, habitat disruption, and, yes, fungal pandemics. I'm not talking sugar maples, a sweet commodity now in steady northward retreat. Or the silver maple that rises tall on the eastern side of the yard. Its canopy begins so high in the sky that all I notice is its thick, furrowed trunk; its branches seem to belong more to heaven than earth. I'm talking *Acer rubrum*, the red, or swamp, maple, as well as its invasive cousin, the Norway maple. This weed of a tree spreads its leaves first and fully and its toe-stubbing, concrete-cracking surface roots far and wide, hogging both sunlight and rainwater. A single tree just one foot in diameter can produce up to 1 million seeds.

Having come up with the neat evolutionary trick of creating countless winged seeds rather than a more modest crop of hard-shelled nuts, maples are unequaled seed dispersers (though this year the sycamores are giving chase). A whirligig can flutter afar and take root in any nook and cranny—a damp gutter, a gravel driveway, between the cracks of a wood-slatted patio. They particularly thrive in rotting wood-chip mulch, and as such would take over the entire yard if given a chance. They already did, once.

Swamp maples are the bane of my backyard, though their leaves do make good fodder for my pile. In fact, I read on the Cornell University Waste Management Institute's website that maple leaves have a 30:1 ratio of carbon to nitrogen, ideal for composting.

Oak leaves, with their higher levels of tannin, have a ratio more like 60:1, which means they take longer to decompose and require more green material high in nitrogen to spur their breaking down.

In the years I've owned my property, I've dispatched many of the maples that had overtaken the landscape, the biggest by tree crews, the saplings grubbed out by shovel, and many more startups plucked by hand. I weed as many maple seedlings as I do dandelions or even crabgrass.

The hardwood forests of eastern North America have changed radically over not just geologic but also generational time. When European settlers arrived, a squirrel could travel from Connecticut to the Mississippi river without ever touching ground—or scarcely a maple tree. The king of the forest was the chestnut, which once amounted to a quarter of all trees in our native forests. It was followed by oak, hickory, and other stout hardwoods, prized for the quality of their wood and the useful fruit from their seeds. Known as the "Sequoia of the East," the American chestnut was undone by a fungal disease brought home from Asia, which killed billions of the trees in the early twentieth century. The stately elm, another native which enjoyed a brief but brilliant reign as the tree of choice for the emerging urban streetscape, was laid low in the 1920s by a microfungus dispersed by bark beetles, also introduced from Asia via Europe. How's this for trading down: Today, *Acer rubrum* is the most widespread tree in eastern North America.

To counter the red maple menace, each fall I empty my pockets of the acorns I've collected from walks in nearby woods, tucking the nuts into the wood chips atop the perennial beds that border the

lawn. Many of these transplants are plucked from the ground by the relentlessly searching squirrels, which have a taste for even sprouted acorns. Bigger oak saplings are munched on by deer. Both animals strike me as short-sighted feeders—their thoughts toward the future end with their next meal at the tips of their noses.

Despite this backyard war of attrition, enough young oaks as well as the beech hedge, two hickory trees, the tulip poplar, and a stand of white pine tucked between the heap and the Tremblays' house have made it through the gauntlet of foragers and other obstacles to rise high enough to begin to take their place in my landscape. One favorite is a mail-order chestnut, a newfangled hybrid said to be resistant to the fungal blight that felled its progenitors.

I tend to these young trees as other gardeners fuss over roses. They will mature long after I'm gone from the property and may well require further care and culling, but I'm proud that my own little niche of a backyard now stands as a nursery and preserve for a modestly diverse collection of old-school native forest, mostly oaks, which sustains an astounding array of native fauna. "The creation of a thousand forests is in one acorn," writes Ralph Waldo Emerson.

I don't always know exactly what kind of oaks I'm bringing home, but judging from the variety of acorns and the leaves on their fledgling limbs, I now have six or seven different types—white, black, red, bur, and what I think is a chestnut oak. I know for sure the names of the three biggest oaks, all now well over twenty feet tall. One is a pin oak salvaged from the clearance-sale section of the garden center when I was shopping for native perennials the

first fall I lived at the house. I took it home for the bargain price of $10, its wispy branches fluttering out the back hatch of my SUV. Charlie Brownish back then, its limbs are still spacious but strong, allowing the sunlight from the western setting sun to filter across the yard.

On the other side of the backyard a red oak has risen up through the inner reaches of the forsythia hedge. Its mother tree grows in the Grissoms' front yard and casts volumes of acorns down onto the T intersection where the streets meet at my corner. Turning cars crush much of this mast, which I sweep up from the gutter to add to my pile. I suppose one acorn jumped the curb and now grows tall in the morning sun.

While huge old oaks still rise from the rocky slope that borders one side of our valley-like road, I see few young ones growing in their shade. Along with deer, squirrels, chipmunks, and other nut lovers, another culprit may be turkeys. Unregulated hunting, which started soon after the Pilgrims' first feast, wiped out the native bird in New England by the Civil War. Reintroduced in the Northeast starting in the 1970s, turkeys are now common but still a treat to see.

A turkey was the first bird Cole saw up close. When he was a toddler just walking, we visited the local nature center, which housed an aviary where various birds are rehabilitated. A gaggle of turkeys was hanging out in the parking lot, acting like they owned the place. Cole cruised right through the flock like it was perfectly normal, eyed by the biggest Tom, taller than he was. Turns out if you take away the guns, turkeys tolerate people pretty well.

As he grew into boyhood, Cole and I often walked up the street to a steep wooded lot where he liked to play make-believe. It is just past the old onion barn turned house, where a faux street sign is posted, "Turkey Xing." The big birds liked to hang out on the shady slope above the road, and Cole would try to sneak up on them. "Of all the paths you take in life, make sure a few of them are dirt," said John Muir. One day we stuffed our pockets with acorns from the lot and, channeling Hansel and Gretel, dropped them along the curb all the way home in hopes the turkeys would follow us back to our bird feeder. They didn't take the bait, even though acorns, particularly those from the white oak, are a favorite food.

A white oak that rises just under the dripline of the sycamore hard by the Rosens' driveway is another prized young tree. One of the reasons I was so quick to call in the tree trimmers last fall was to have them remove a long, low sycamore branch growing perpendicular to the property line, threatening the white oak's future prospects. I know it's a white oak because it's one of the few types of deciduous trees that hold on to their faded leaves all winter, a phenomenon called marcescence. That screen of wide, dangling dead leaves helps shield the view of the Rosens' driveway through the winter, and when the fat, wide leaves do drop onto the bare ground of the mulched garden beds, I know the new growing season is just around the corner. I'm happy to add these latecomer leaves to my pile, especially at a time when old brown matter is in short supply to balance out the first cut of springtime grass.

"A yard without oaks is a yard meeting only a fraction of its life-support potential," I read in Douglas W. Tallamy's *The Nature of Oaks*. Oaks harbor more life forms than any other North American tree, including hundreds of kinds of caterpillars. That may explain why for the past several seasons a robin has made its nest in the crook of a branch and the smooth, straight trunk, about ten feet off the ground. I spotted three bright blue eggs in the nest a week or so ago when I leaned a ladder against the already stout trunk, using the extension saw to prune a limb from the top of the tree that dared point toward the Rosens' house.

> "If we are going to design landscapes that enhance local ecosystems rather than degrade them, we need to include oaks because they are the best tree genus in North America for supporting the food web that supports birds, the insects they eat and thousands of other species. They also capture more carbon than other trees and hold more rainwater on our properties than other plants." – Douglas W. Tallamy

These young oaks may well grow on this piece of ground for the next 100 years—swamp maples rot out after about fifty years, often ending up on a roof or across a power line. Though I use the mower to mulch up countless samaras that fall across the lawn, I've never trusted adding the seeds wholesale to my pile at this late point in its seasonal cycle. I've always feared too many would survive.

So each May I sweep and blow away enough maple whirligigs from the porch and driveway and out of the gutters to top off the small

plastic tarp two times over. And though I'm sure the seeds are loaded with nutrients, I dump them next to the brush pile under the old silver maple in the corner of the yard. It, like two of my other remaining maples, is technically on town property—safe from my attention for now. I let the seedlings rot at the base of its trunk.

All these maple seeds don't deserve my pile, is what I'm saying.

JUNE

Green Machine

Though this half of the planet still tilts toward spring, the calendar says June and the vibe says summer. The onset of hot, sunny days and warm, humid nights, along with a thunderstorm that spoiled the Memorial Day parade and drenched the backyard with an inch of needed rain, have conspired to send the lawn into overdrive.

The early-flowering bulbs, azaleas, and bleeding hearts have prospered, peaked, and faded. Until the summer flowers arrive, the color of my backyard is leafy green, and that's money for my pile.

A 2005 NASA study estimated the area covered by lawns in the United States to be about 63,000 square miles, making it the nation's largest irrigated crop by area, three times the area of irrigated corn. Today, four out of five American homes have grass lawns, with lawn care now a $40 billion industry.

That's a lot of green, and all my share stays in my back pocket and backyard compost pile. In *The Lawn: A History of an American Obsession*, Virginia Scott Jenkins traces the historic desire, transplanted from the landed gentry of Europe, of Americans to feature squares of *tapis vert,* or "green carpet," as part of their property. Though borrowed from abroad, the notion is now as American as George Washington and Thomas Jefferson. Both founding fathers grew English-style lawns on their colonial estates, kept close cropped by resident sheep, horses, cows, and, no doubt, the labor of slaves.

With the development of suburban housing and Americans' mass exodus from cities, not to mention the invention of the lawn mower in 1830, "lawns became emblems of American leisure and prosperity, a symbol of man's control of, or superiority over, his environment," Jenkins declares.

Our shared obsession with lawns is fertile ground for anthropologists like Krystal D'Costa. Writing for *Scientific American*, she asks why Americans place so much importance on lawn maintenance, then helpfully answers her own question. My grass-stained sneakers and I feel seen:

"The state of a homeowner's lawn is important in relation to their status within the community and to the status of the community at large. Lawns connect neighbors and neighborhoods; they're viewed as an indicator of socio-economic character, which translates into property- and resale values. Lawns are indicative of success; they are a physical manifestation of the American Dream of homeownership." – Krystal D'Costa

I have family roots, as it were, in growing grass. After my uncle lost the family dairy farm in Nebraska, he moved to Colorado, where he became the manager of a large turf farm on the outskirts of Denver. I visited Uncle Jerry once and he led me on a tour of the facility, which comprised vast, flat acres of grass growing under the hot sun of the high plains and rainbow sprays of pivot irrigators. He was proud of the work he did, which was to produce ready-made lawns for homes across the Front Range and up to the vacation estates of Aspen and Vail. His sod graced the second home of former President Ford, he told me that day, and he was just as proud to mention he grew the turf laid down at Mile High Stadium. (That turf, 100 percent Kentucky bluegrass, is kept green year-round by twenty-one miles of heated pipes laid down underneath it.)

The machines that scalped thin strips of sod from the ground to roll them into tight bundles, roots outward, were impressive, but what struck me, walking along the edge of the green fields, was the lushness of life in the sagebrush prairie just beyond the reach of the sprinklers. Each step produced a ripple of fat green grasshoppers staying a jump ahead of my feet. Big, rangy jackrabbits zigzagged away on our approach. The dusty roan dirt was pockmarked with burrows, from gophers to ground owls. We flushed a covey of quail and surprised several rattlesnakes, who shook their tails at us. I don't recall seeing a single living thing on the bright green turf itself, but all manner of wildlife thrived along the untrammeled margins of the sod farm.

While I take every opportunity to remove what turf I have to expand the garden beds that ring the house, patio, and property, I admit to a lingering lawn obsession of my own. I see my patch

of turf as a boon for the true object of my backyard affections. Another factoid from the web: The average half-acre lawn in New England—only slightly larger than my property—produces over three tons of grass clippings a year. Like I said, that's a lot of green, and it's too good to go to waste.

As voluminous and valuable as grass-cycling is to my pile and me, it's not even close to being the most recycled product in America. Know what is?

Asphalt.

This fact came to light when I came home early from work to find a crew in the process of stripping the street along my property of its old and crumbly asphalt veneer with a huge mechanical contraption that grinds the pavement into pebbly chunks and sluices them up a conveyor belt into a waiting dump truck sidled alongside, a procedure not unlike my uncle's strip mining of sod. If the street crews are busy blacktopping the nation's roads late in the day, it truly must be summer.

I've let the lawn go through the bulk of No Mow May and past Memorial Day weekend in favor of beaches to picnic on before the storms set in. Now the bill has come due. The grass is tall and thick enough not just to swallow the tennis balls I toss to the dog but also to hide his morning poo, which is my marker for high time to mow. But with the din of the repaving, I head inside to wait for the machines to make their way further down the street. It's hard to decompress in gardening mode while a lumbering machine is beeping and belching and spewing asphalt on the other side of the forsythia hedge.

Joni Mitchell's siren song had it right. We've paved paradise: According to Yahoo, "The United States has paved 3.9 million miles of roads, enough to circle the Earth at the equator 157 times." Another jaw-dropping stat that hits close to home: There are some 2 billion parking spots in the U.S. alone, enough to pave over the entire state of Connecticut.

As environmentalist Rupert Cutler noted, "Asphalt is the land's last crop." Unless you count lawn grass, which could cover the state of Texas.

For better and worse, the lawn is my green pavement, and for as long as I intend to keep any of it, I'll try to make tending turf as productive and hospitable to nature as I can. The feedback loop that is my compost heap allows me, at least in this clime, to eschew the chemicals and copious amounts of water that are required to grow enviable sod. "Because much of the country is not hospitable to turfgrasses—none of which are native species—we use 90 million pounds of fertilizer and 78 million pounds of pesticides annually just to keep lawns thriving, bright green, and bug-free," I read in "Blades of Glory," published by *The Week*. These statistics support activist Michael Pollan's own conclusion: "We have begun to recognize that we are poisoning ourselves with our lawns."

Once the road crew decamps for the day, I haul out the mower. The yard is dotted with white buttons of clover flowers. It seems a shame to decapitate them, so I skirt the Toro around the particularly chewy patches to create some new pollinator meadows. Many others escape the high blade and more will soon sprout to nourish both my soil, with their nitrogen-fixing roots, and the buzzing bees,

with their energy-rich nectar. "A carpet all alive," in the words of the poet William Wordsworth.

Honeybees—an import from Europe—and the hazards they face get all the headlines, but there are hundreds of different kinds of wild bees native to just this area. None sting or make honey, nor even hives, preferring solitary nests in burrows underground or in rotting wood. They are vital pollinators, more than butterflies or honeybees, truth be known. A 2023 study by researchers at the University of California found that flowers pollinated by native bees produce more diverse, healthier offspring than flowers pollinated by honeybees.

I've gotten to know a number of species, if not by name—the iridescent ones are called sweat bees—then by the territory they patrol. A big, lumbering American bumblebee owns my vegetable garden; he and his kind nest in the two-by-fours that frame the old open-sided shed that sits on a narrow cement pad in the back of the garden, against the rear wall of the house. Years ago, it was home to the fuel-oil tank, but after switching to gas I had the tank removed and put in a potting bench. Cole and I once used the shed's slanted roof to get up onto the roof of the house, usually to retrieve a wayward Frisbee. That perch is no longer safe, as the bumblebees are in a race with the carpenter ants to see who can bring down the old lean-to first. I don't really mind; I hardly use the shelter anyway, and the bumblebee needs all the help it can get, having lost 89 percent of its population over the last twenty years, a rate that outstrips even the decline of the equally iconic monarch.

Fireflies are also in trouble. "Blink and you'll miss them this summer. Around the world, people are reporting that local firefly

populations are shrinking or even disappearing," notes John R. Platt in an article on takepart.com. And fireflies do much more than light up summer evenings with their phosphorescent glow.

"The loss of fireflies, which are beetles, can have multiple effects on their ecosystems," says Ben Pfeiffer, founder of Firefly.org. "For one thing, some firefly species—there are at least 170 in the United States—play a role in pollination. More important, however, firefly larvae are voracious predators that live in the ground and eat slugs, snails, worms, aphids, and other problem critters that would otherwise grow out of control. I call them nature's pest control."

Having skirted the patches of clover, I mow around the mini meadows of uncut grass. The grass ripening to hay stands out like a bouquet on a tabletop and makes the cut grass that surrounds it look all the more manicured. At least to my eye; Carl has told me some other neighbors have been wondering what's up with my yard and its rewilded look. But I like how, especially at dawn or dusk, the slanting sunlight passes through their stems and highlights the seedheads. A vision of amber waves of grain is deeply ingrained in the American psyche. I imagine the fireflies as well appreciate not having a mower running roughshod over their turf; in the years since the grub wars, my yard has become the go-to place for the neighborhood kids to chase after lightning bugs each July. I will cut the mini-meadows by late summer (after the clover flowers have faded and I've harvested a few fistfuls of grass seed), but for now they stand as symbols, exclamation points, really, of pastoral nature left to be.

I have mostly white clover in my backyard, though there is some more gangly red here and there. A member of the leguminous pea

family *Fabaceae*, clover is cultivated around the world as nutritious fodder for livestock, a source for sweetly flavored honey, and a cover crop that naturally fertilizes the soil. Anyone with a bit of Irish in them, like me, loves their shamrocks, and as a kid I happily hunted for four-leaf clovers.

The nitrogen-fixing bacteria that grow in symbiotic partnership with clover roots convert atmospheric nitrogen into a form that nourishes the plants and others around it. When clover dies, the fixed nitrogen is released, fertilizing the soil to the tune of up to 200 pounds per acre. And that's why both my lawn and pile are in clover.

Soaking It All In

A dry spell, classified as a moderate drought by the local weather report, draws me outside after a day at work in my office cubicle. I relish chasing the daylight with a walkabout in the backyard. My evening perambulations allow me to track how the various living things I tend to are developing from day to day. It's the time of year for such pursuits. I see the new tomato plants and salad seedlings need watering. My pile looks a bit thirsty, too.

Often at this time of year, without much of a garden agenda and with few pressing yardscape duties, I simply take a seat and take in the remains of the day. A small plastic chair on the back-porch stoop puts me right at eye level with the top of my pile's log walls, so I can easily gauge its waxing and waning from afar. As it reduces under the summer sun, the heap levels out with the top edge of the chicken-wire fence along the back, allowing me to eyeball its shrinkage against the steel-gauge grid. At this accelerated stage in its lifecycle, even a pile fluffed full of seaweed scraps, grass clippings, and a week of food waste can flatten by a full wire row on successive days—faster if it rains.

"Nature does not hurry, yet everything gets accomplished." – Lao Tzu

One of the charms of southern New England in June is just how long the days are; sunset tonight is well after 8 p.m. Better yet, with the lack of rain comes the bonus of enjoying the backyard at twilight without needing to mow the lawn—or having to swat away pesky mosquitoes.

In this well-watered climate, kept cool by the nearby Long Island Sound, I normally don't have to water my garden until the dog days of August. The storms that sweep up the Atlantic seaboard or rake eastward from Tornado Alley in the Plains States deliver frequent dousings, whether in the form of snowflakes, misty showers, or full-on gully washers.

Or, as these past couple weeks have shown, not. Though a stretch of showers is predicted to arrive later in the week, I fret that if I don't soon water the seedlings in the vegetable garden and the recent transplants among the perennial beds, all my care up to this point will be wasted. I take particular responsibility for a chestnut oakling that sprouted in the vegetable patch, which I've moved to a spot in the wood-chip mulch along the back fence. The oak might live for the next century—if it can make it through the next few days.

Taking my leave of the backyard bench, I lift the hose from its perch against the house and stretch its loopy coils alongside the vegetable garden. I enjoy hand-watering, the feel of a thumb growing numb with the gush of water chilled by the buried pipes that deliver it from a reservoir twenty miles upstream.

Watering with the hose gives me control, a say in what grows. And in addition to being stingy, I like the feedback. I trace the spray

around the dripline of the oakling, trying to envision how the fluid will work its way into the ground that surrounds its fragile young roots. Am I helping give them a go, or drowning them?

In the vegetable garden, I feather the water in my best imitation of a gentle, soaking rain. I keep a couple old five-gallon paint buckets near the back door and fill them now too—for the dog, as well as spot watering. In spare moments throughout the week, I'll haul a bucket or two over to transplants that need it.

I stretch the hose out to the back corner of the yard and water the shade garden. Hard by my pile, it's the lowest point on the property, to and through which all the water drains. I fancy it as my wetlands and have planted it as such, with an array of ferns and lilies and even a rescued skunk cabbage. A swamp maple in a nearby preserve had blown over in a storm, and the fragrant native was dangling high and dry from the fringes of the tree's upturned root ball, its fate all but sealed. I plucked the plant from its perch and took it home to the lowest vernal spot of the yard, where it's since thrived.

In the center of this patch, on a tiny hummock of high ground, grows a twin-trunk paper birch, brought home years ago as two twigs in cellophane wrapper, a giveaway from the town on Arbor Day. After the growing stems got munched the following year by deer, I surrounded them with a stretch of bird netting wrapped around spikes of bamboo, which I cut from twenty-foot-tall stalks of a patch of the pernicious invasive that grows on an untended strip of land alongside a nearby road. This past spring, the wrap finally came off; the birch is now head high, its papery trunks beginning to peel in fine fashion. This back corner of the yard is always next-to-last on

my route with the hose, and I admit to spoiling it like a favored child with extra doses of water during dry spells.

Finally, I stick the nozzle end into my pile, first along one of the log walls, then the other. I can't see where and how the water flows into the mix but figure the two flanking sides of the heap have been disturbed the least over winter and early spring and are likely still mostly a mix of leaves and compost pulled from the center during past excavations. Surely these sections could use a soaking to help them catch up to the steaming mix of grass clippings and kitchen waste cooking away in the cauldron that is the center.

"The average American family uses 320 gallons of water per day, about 30 percent of which is devoted to outdoor uses," states the EPA's WaterSense website. "More than half of that outdoor water is used for watering lawns and gardens."

I do feel rather foolish about dousing a pile of rotting leaves, but when it comes to watering the lawn, I am all about tough love. "Lawns consume massive amounts of water," I read on scienceline.org. "There are forty million acres of turf grass in the United States, covering 1.9 percent of the land. If all that is kept well-watered, it could use sixty million acre-feet of water a year. That's more water than is used to grow all the corn, rice, and alfalfa in the country, as well as to irrigate every orchard and vineyard."

How best to irrigate a suburban property while minimizing the waste of water is a touchy subject. Hand-watering versus an in-ground sprinkler system? That question is akin to the argument of using a rake versus a leaf blower. I'm no Luddite and see the value

and ingenuity of, say, using a high-tech drip system in a more arid garden—and the impracticality of setting up the requisite network of pipes and valves in this hardscrabble yet well-watered clime. Mostly I prefer the simplest modes of gardening—using a hand rake or thumb on hose when I can. It's peaceful and sustainable. The backyard, and my pile, is a refuge from the remote and abstract technology and material consumption that now define our lives, and my manual labor keeps me in touch with a natural, hands-on state of being.

Brush Off

It's hump day, a Wednesday, yes, and also the summer solstice. The Earth's axis is now tilted as far as it can go toward the sun, which makes it hump day for the whole year. As far as daylight goes, it's all downhill from here until we pass the shortest, darkest day of December.

I make hay of the sunshine by tackling a recurrent backyard chore: pruning the bountiful growth of the shrubs, bushes, and small trees that populate my property. Properly spaced and prudently tended, these plantings make up the bones of the classic suburban landscape. Among other attributes, they create a beautiful, shady backdrop for flowering annuals and perennials, provide screening from the street and neighbors, and offer a wealth of food and shelter for all kinds of backyard birds and critters. Left unchecked, this understory of big bushes and small trees would, in a few short years, overtake the property, returning it to the state in which I found it years ago—an uncivilized briar patch of brambles and brush and vines, all scrambling to crowd each other out.

This dynamic—some would say romantic—tension between chaos and cultivation was perhaps best expressed by Vita Sackville-West. As Sarah Raven writes in *Sissinghurst*, a biography of her mother-in-law and the garden she created from the ruins

of an Elizabethan estate in the south of England, "Vita loved her borders to be packed. She hated the sight of too much mulch, criticizing Edwardian rose gardens with their 'savagely pruned roses of uniform height, with bare ground in between, liberally disfigured by mulches of unsightly and unsavoury manure.'

"An enchanting garden like Sissinghurst is, I would say, at its most beautiful at precisely the point where its informality is about to tip over into chaos. I am with Vita and her desire for *sprezzatura*—a studied nonchalance, a balance of formality of structure with informality of planting."

I'm with Vita too, save for her dissing mulch and manure. So tonight I browse the yard with clippers in hand to help guide the growth of shrubs and bushes, tidying up the property and freeing up time for the coming weekend, which culminates in Father's Day.

Tending a backyard, garden, or compost pile is very much a parental exercise. My goal is to raise a yard that is sturdy, strong, and resilient; one that's a product of its environment, yet individual, a unique creation I can be proud of having nurtured, one that is not a danger to itself or others.

Just the other day Cole was proudly showing off the scars he's collected over the years—one a lasting scrape on his kneecap from falling off his bike, another a puncture wound on his calf from trying to jump over a neighbor's oversized hydrangea while playing tag with an older boy, and a nick on his forearm from when he rounded the backside of the compost heap and snagged himself on a rusty piece of chicken wire. This last incident led to a quick trip to the

doctor for a tetanus booster and a return, by me, to the heap with a pair of wire cutters to snip the offending piece of fencing.

The one tree I've hardly touched is the most memorable on the property. When first grubbing out the widow's backyard, I saved just one sapling from the thicket I cleared to make the lawn. With its straight trunk, smooth skin, and serrated compound leaves, I could tell it was a hickory, a hardwood unlike the trashy swamp maples, and unhindered by vines. The four-foot tree was just as tall as my son; from its branching out every ten inches or so with stubby limbs, I figured it was about the same age. I left the little hickory in place and grassed around it.

I've never pruned the tree, save for clipping two of the lowest limbs pointing toward the house, at shoulder level, to guide their growth upward and out of eye-poking range. Otherwise, the stripling rises straight and true, fully circumscribed by branches from top to bottom. Stiff and shock resistant, hickory wood is ideal for tool handles, wheel spokes, and drumsticks. Golfers long played with "hickory sticks." The Ojibwe name for hickory translates roughly to "hardwood bow"; the word itself is Algonquian. I prize hickory firewood, though it's tough to split, and my father never grilled steaks without first soaking hickory chips in water and tossing them on the coals to add flavor.

Last week, just before his high-school graduation, Cole was ambling about the backyard and noticed our hickory, as if for the first time. I shared the tree's history, and he walked over to encircle the trunk with outstretched hands. He could barely touch thumbs to fingers. Straddling its base with his feet, Cole leaned

back to peer up to the crown, forty feet above. "Wow, it's gotten big," he marveled.

So this much is confirmed: I have raised a tree-hugger. The further adage about acorns falling from trees also applies, and the moment made a good enough selfie to send to Gramps on Father's Day. The hickory's bearing and beauty suggest to me that it will remain standing whatever happens to the house. I wonder in the years to come, long after my pile and I are gone, whether Cole will drive by with his own kids, point to the tree, and tell them its story.

Like parenting, pruning is idiosyncratic, yet it generally follows fairly well-established guidelines in terms of where to nip, how much, and when. I start tonight's session with one of the more straightforward pruning tasks—cutting the old-wood stems from the hydrangeas that bloom big and bold each summer. The bases of the plants—I have two in the perennial garden—are thick with fat green leaves emerging from the bloom buds. But rising from each plant are dozens of old-wood stems from last season, a thicket of stout, three-foot-tall hollow shafts with dagger-like ends where I deadheaded the big blue flowers last fall—I cut them back in deference to any future games of tag.

Though I prune with some impunity, to tidy things up around the edges or remove a limb hanging over the house or power line, the backyard remains shaped in its own image. It's still thick with non-natives that were planted or took root long ago and that, through pluck and perseverance, have become permanent residents of the local landscape. The toxic yew and fragrant japonicus alongside the front of the house, likely planted by the first homeowners seventy

years ago, have grown thick and massive. Each time I prune their tops and the side branches that scrape against the siding, I have to take care not to disturb the two or three bird nests hidden within. Mostly, these foreign ornamentals remain because as I remove lawn and expand the flower beds with each passing season, I create plenty of new spaces to bring natives into the mix. These naturalized citizens of the backyard are especially profligate growers. The old lilac that grows outside the back door requires reining in or else will flop over with the weight and drama of its seasonal bloom. Early each spring I snip the flowering tops to capture their aromas in a vase indoors, but still it grows scraggly. I prune best just after a summer rain, when the wind and water expose the weakest and most wayward of new growth. These gangly branches bow, foreshadowing their fate, whether to be sheared off by an August squall or snapped under the frozen weight of a February ice storm. Or by my hand. A thundershower passed earlier in the week, causing a main branch of the lilac to droop over the limestone steps that lead to the kitchen door. I lop off the limb and add the lot to the brush pile. Hard to say who I pick up after more, my son or the bushes of the backyard.

My backyard is no poodleized topiary garden. As with Vita and with Cole, I give the plantings a lot of leeway, allowing them to grow on the wild side while trying to maintain some sense of order. Put another way, my gardening ideology is a mash-up of "Father Knows Best" and any of the more modern dimwit-dad sitcoms. It's a nature vs. nurture struggle for control that has played out over the centuries, as Eleanora Montuschi, a professor at the London School of Economics and Political Science, describes:

"The image of man's dominion over nature is deeply rooted in Western thought. It first appears, in different forms, in the Book of Genesis. It also reappears as one of the leading images of the emerging 'new science' in the 16th century. Starting from the 17th century, gardens were designed as geometrical spaces. Plants and bushes were cut into triangular, spherical, conical, and pyramidical forms. Sometimes they were shaped as animals or human beings. In other words, nature was altered by imposing specific forms over her spontaneous ways of expression.

Instead, in the 18th and 19th centuries, gardens were conceived in view of complying with nature. Nature was to be allowed to express herself in her own forms: she was at most to be 'perfected.'" – Eleanora Montuschi

That about sums up the duality of gardening and parenting: striking the balance between being a control freak to maintain order and letting things develop with a gentler guiding hand.

Nature, like a child, will always find a way to express itself, fully in "her own forms" and in keeping with the soil in which its offspring is nurtured and under the sun that shines upon them. The day I asked my son, "Do you want me to tell you what to do, or do you want me to let you make your own mistakes?" and he answered, "I'll make my own mistakes" was the day I realized nature was taking over nurture. And everything (I don't recall what, exactly,

-the-making we were talking about) turned out just

Pruning is about guiding growth in the right way, with a fair amount of culling, in a never-ending pursuit for some sense of order and control. It's also about knowing what not to cut. An ill-treated plant will, in time, let you know when you've done it wrong. Sometimes Father doesn't know best. Pruning, and parenting, are about letting go and recognizing that not everything will fit into a neat package, or my pile. The tangled stack of brush I leave at the side of the curb is ample evidence of that.

But both child and garden tend to work out as long as you keep serious problems from taking root, prune when necessary to help guide growth, ensure sufficient food and water—and always add as much compost as you can.

Scratching the Surface

It's a Friday in late June, a perfect midsummer's eve for some leisurely yardkeeping. In my world, cocktail hour is garden hour, and unwinding from the workweek by reconnecting with the ground I keep is a pleasure, especially at this time of year when the plants are intoxicated with new growth.

I start by mowing the lawn, collecting three hoppers of grass clippings. Once again, I skirt around the small island meadows left to grow in the middle of the lawn. The uncut stalks of grass have turned tawny brown and bow their seedheads with the breeze. I spot a bee or two nosing about the fading clover flowers and, nearby, a few butterflies hovering around the garden to help with pollinating duties. Tuffy the cat has taken to lounging in these meadows. A few evenings ago I watched as he rose up to pounce back into the grass like a fox. He'd flushed a vole, who then scampered across the cut grass to the next meadow. I was so impressed with the vole's great escape that I scooped up Tuffy from his island catio and brought him inside.

The unmowed meadows are imprinting themselves on the landscape. One spreads like an apron from the stump of a large swamp maple in the back corner of the yard. The grasses and goldenrod rise the tallest and thickest here, in the lowest, wettest

part of the lawn. I had asked Chris the woodman to leave the stump knee high, and I used it as a seat until the smooth ringed surface collapsed with rot. After a favorite old pitchfork finally fell apart on me, I stuck the jagged end of the broken shaft into the spongy heartwood, a Connecticut gardener's version of King Arthur's sword. Birds often perch on the rusted metal joins of the handle grip, which likely explains the raspberry vine that now flowers from the stump's crumbled flank. So I have to mow around the spiky branches anyway.

The meadow not far away is crescent shaped, following the contour of the flower garden that curves around the corner of the house. Its placement also happens to be the geometric end of the pattern in which I typically cut the grass. I'm happy to leave these last steps of the mower not taken, as this prairie moon is now thick with clover and ripening stalks of rye, plus one very lucky vole.

As I put the mower away and empty the last hopper of clippings, I consider my pile. Over our last few sessions, I've thoroughly worked its front and back sides, strip-mining the flanks to aerate it and mix it full of fresh greens from the yard and kitchen and dried brown leaves gleaned from the corners. It's time to shift my pile from side to side.

Having gouged out the periphery of the bell-shaped heap for crumbly leaf mold, I've exposed the four corners of the log walls inside the crib, leaving a reach of only a couple feet or so in the lower middle untouched. I start in on the left side, channeling my way along the wall from front to back with the hay pitchfork. The deeper I drill, the richer the compost. My pile feeds off the decay of

the log walls that contain it, themselves well on their way to rotting. I pry out thick wads of raw leaves from the crevasses, stuck like dreck between teeth, which will be perfect for mixing with today's grass clippings.

I've created a foot-wide gap along the length of the heap and portside log wall. The entire left side is revealed in cross section, a thick stack of cold-pressed compost that begs to be further teased and turned out. This Humpty-Dumpty of a pile needs a great fall.

To gain ready access to this newly exposed quadrant of the heap, I teeter to their sides the middle two logs. Standing perpendicular to the wall of newly exposed compost, I step in close with the pitchfork to unbind the pressed leaf mold and seaweed laid down early last fall and turn it loose onto the top and backside, mixing the oldest part of my pile with the newest.

At last I've chiseled deep enough into the bottom so the overhanging matrix tumbles to the ground. I set the two logs back in place and draw some of the newly expunged leaf mold up against them, adding at intervals fresh grass clippings and tossed layers of rotted leaves from the backside. In all, I've carved about a third of the way into the left side. My pile is now a slightly lopsided version of its former self, taller and much suffused with air and freshly mixed compostables. It turns out you can put Humpty-Dumpty back together again.

The next morning, the heap is in its cups, positively Falstaffian in its fulsomeness. I'm surprised at how little it has settled, even after an overnight rain. The leaf litter across its rounded surface is moist and crumbly, mottled clumps of caked-together leaves bound with

layers of humus in the making. The whole lot is well on its way to recomposing itself as a matrix of new earth that I can shovel up instead of disassembling with pitchfork.

My main chore today requires a different kind of dirty work. The street that passes along the front of my house has a gutter. But around the corner the gutter ends, and the rainwater continues ever so slightly downhill along the forsythia hedgerows toward the storm drain down the street.

The slowing runoff, thick with silt and dander from the trees that overhang the street, deposits itself along the forsythia bushes and proceeds to blossom in a long line of weeds and maple sproutings. So I take the squared-off flat shovel to plane away the thin layer of alluvial soil and its crop-top of weeds. I drag the small tarp along the way, and in just a few minutes I reclaim a good foot of roadway for a stretch of sixty feet or so, procuring several tarpfuls of moist, loamy delta dirt. I feel like I'm tilling the banks of the Nile. Better to deposit the fill in my heap than let it pile up along the street or clog the storm drains.

Adding soil to my pile may seem like bringing coals to Newcastle, but dirt ain't cheap if you have to buy it, and the billions of microbial creatures within each shovelful will further fuel the composting process. I am sure the detritivores of the heap will subsume and supercharge the newfound weedy dirt into something not just clean but with a little extra oomph in it.

These days, especially when my pile is damp with rain, it takes only a scrape with a rake or pitchfork to uncover all manner of millipedes, roly-polys, skinny red worms, and fat racers. It's amazing to see how much life is contained within the heap, even just scratching the surface.

A review of existing research published this summer finds that soil is likely home to nearly two-thirds of all life—everything from microbes to mammals—making it the singular most biodiverse habitat on Earth, I read in *The Guardian*. This includes fully 90 percent of fungi, 86 percent of plants, and 40 percent of bacteria. "The actual figure could be even higher as soils are so understudied," the study's authors claim.

"Scientists estimate that only 10 percent of small soil animals have so far been identified. We know even less about their relationships," George Monbiot says in his 2022 book *Regenesis*, about how to transform the food system, in part by restoring the soil.

"Beneath our feet is an ecosystem so astonishing that it tests the limits of our imagination. It's as diverse as a rainforest or a coral reef. We depend on it for 99 percent of our food, yet we scarcely know it. Leonardo da Vinci remarked that we know more about the movement of the celestial bodies than about the soil on our own planet."
– George Monbiot

Monbiot is discussing soil ecology, but the point holds true for my pile, perhaps even more so. He explains: "The soil might be the most complex of all living systems. Yet we treat it like dirt." He

likens soil to a wasps' nest or a beaver dam: a system built by living creatures to secure their survival. "Microbes create aggregates by sticking tiny particles together with the carbon-based polymers, or cements, they excrete. In doing so, they stabilize the soil and assemble habitats for themselves. Over time, this process builds an ever more complex architecture: pores and passages through which water, oxygen and nutrients pass."

Each morning and evening when I let Miller out the back door, he makes a beeline for the back corner of the yard and circles around my pile. Sometimes he chases off a squirrel or sniffs out a chipmunk, but more often on these midsummer days, morning or night, he gives flight to two or three robins that have taken to perching atop the heap. I can see telltale signs they flick about the surface, scattering flecks of leaf litter in search of easy pickings.

Not unlike the peckish birds, I inspect the surface of my pile each time I visit, plucking out the twigs and wood chips that continually bob up like corks to the surface. I seldom notice these woody chunks and stems when I rake the leaves from the yard each fall. Most of the chips get hoovered up when I run the mower along the mulched garden beds. Some twigs fall directly from the overhanging maple that shades my pile, but all these stray pieces remind me of how much the trees in my yard continually crop themselves, shedding not just leaves every fall but spent flowers and seeds and bits of branches and limbs and bark all along the way, which fall to the ground to be broken down further by hand or blade. Trees are gloriously messy things.

Each twig or chip I toss aside is one less piece I'll need to screen from the finished compost when it comes time to disperse it

across my lawn. Well, not so much need as should—screening is a laborious task I rarely bother with anymore. Mostly such grooming is my way to stay connected with my pile, much in the same way I enjoy picking up seashells and curious rocks at the beach or even checking the dog for ticks. A touch is sometimes all it takes to stay connected, as any chimp will tell you.

Sometimes, a bright red male cardinal will alight on the pile and chase away the robins. Though his frugivore beak is specialized for cracking seeds, cardinals have a bit of the blue jay in them, so he could also be tromping about for a beetle or grub (though lord knows there's plenty of seeds to be found in the heap). I found one once in the Havaheart trap I'd set out to catch the jumping mouse; he was lured by the schmear of peanut butter bait. "Cardinals appear when angels are near," the saying goes. Redbirds are always a welcome presence in my yard.

Today, the roly-polys are out in force, and I linger to follow one traversing the escarpment of crumbled leaves. Roly-polys have fascinated me since I was a kid, in much the same way sightings of horseshoe crabs did on summer vacations to the seashore. Both appear—heck, truly are—prehistoric, having persisted since before the dinosaurs were even a thing. I like how people call them whatever they want: doodle bugs, woodlice, pill bugs; Brits know them as chiggypigs, penny sows, and cheesybugs. I've always thought of them as mini armadillos. Whatever name you call them, they were made for composting. Thanks to a special brand of bacteria in their gut, roly-polys are incredibly efficient at chewing up rotting vegetation. Their diet includes self-coprophagy, which means they scarf up their own feces to take in nutrients they may

have missed the first time around. Being tolerant of ammonia gas, they don't urinate, and instead excrete waste fluids through their shells. They're such tough little guys, they can take in heavy-metal ions and crystallize them in their bodies, which is why researchers are studying how they can be used to help detoxify contaminated industrial sites.

And guess what? Like my earthworm buddies, the roly-poly isn't native, having rolled up to these parts from the Mediterranean in the eighteenth century to find its niche in the New World ecosystem. My favorite factoid about the doodle bug? They have gills and are actually crustaceans. And horseshoe crabs are not crustaceans; they are in the spider family. Darwin's evolutionary tree has some really twisted branches. (As it happens, even the earthworm is now facing a fight from a newer colonist, a worm from Asia that grows to six inches and is so hyper in its movements that it's known as the jumping worm. It was imported in the 1940s from Japan to the U.S. by the Bronx Zoo to feed duck-billed platypuses.)

The hard-pounding rain has exposed, as it always does, a fresh smattering of wood chips and twigs to flick away. Flotsam and jetsam from the beach constantly reveal themselves as well—I collect the rubber heel of a flip-flop, a stretchy wristband, and other scraps of plastic to take back inside to the garbage can in the kitchen. Today's surprise find is a salad fork, likely discarded from a dinner plate hastily scraped off by one of the Tremblay girls and dumped into the waste bucket. Its tines are crusted with humus; I'll stick it in the dishwasher before returning it, once again shiny and stainless. My pile is nearly done. I can tell not by sticking a fork in it, but by pulling one out.

JULY

Fire Works

I get off work early this afternoon. It's a getaway Friday for the Fourth of July holiday weekend. I'm hoping to take an overdue walk at the beach in advance of the next night's big fireworks display. Launched from barges moored just offshore from the town's biggest beach, the public pyrotechnics attract thousands of people each year.

Though the long, hot days of summer have slowed the growth of grass in my yard, they've also warmed the shallow waters of the nearby Sound, leaving the local beaches awash with seaweed. I know from years past that the town tidies up the beach before the fireworks by dragging a mechanical sifter across the sand. I hope I'm not too late to do some beachcombing of my own. I don't know how much more fresh hot greens my pile needs, but I like the idea of finishing it off as I started it—with "the best fertilizer there is."

I have pride of place, living so close to the ocean. But I imagine that for inland composters there are plenty of pond owners and homeowner associations with small lakes who would gladly entertain the idea of skimming off mats of duckweed, fishing out a few buckets of invasive water plants, or mucking out some overgrown cattail to truck back home to a countrified compost

heap. Though certain types of algae turn into toxic pond scum, the sources I've consulted say most freshwater muck makes a fine amendment as long as you layer the compost correctly. (Unfortunately, the oceanful of sargassum that now regularly befouls Caribbean and Florida beaches has been found to have too much arsenic and cadmium—heavy metals that can be toxic to humans—to be useful as compost without treatment. Darn!)

Returning home from the beach with plenty of daylight left, I see that Carl has mowed his grass in preparation for his family's Fourth of July party and left the clippings at the base of my pile. I set about my evening chores to mix in the grass and the seaweed, along with a week's worth of kitchen scraps. Having excavated the left side of my pile a week ago, it's time to do the same on the flip side.

As I prepare to dig in, Danute comes by with a special request that makes the job easier. After several years of talking about creating a garden in her own backyard for herbs and spring produce, she has finally done it. She walks me to the south side of her house, where she has dug up the ground along where the fireplace chimney, once painted white, rises from the foundation. A small pile of unearthed detritus includes clay pot shards and a pair of horseshoes crusted with rust. No doubt the play set had been tossed aside many years ago, then lost to weeds and time.

She has already planted tidy rows of baby Swiss chard, chives, and spinach, and nearby are various pots containing herbs—I see sage, thyme, and rosemary—which she wants to plant in the freshly dug ground. Though I worry about the faded white chips littering the soil, likely from the lead paint era, the little patch, set along the sunny side of the house, has the makings of a fine little herb garden.

What the sunken, bare patch of ground needs now is some bulk, a generous top dressing of compost.

It's only fair: For years my pile has thrived on voluminous scraps from her home cooking, as well as raked leaves from her bare-swept front yard and my freelanced cleanups in the back. In return, Danute has an open invitation to gather fresh pickings from my garden and does so on a near daily basis. But until now, aside from a few clay pots to fill, she's never had need for what my pile produces.

Luckily, I now have compost in spades. I park the wheelbarrow in front of the heap before plunging into the right front corner with the manure fork, turning out cascades of mature compost from a dark, rich seam. I use the spade to fill up the wheelbarrow—the only screening I do is to flick a few clumps of pressed leaves back onto the top of the heap. I trundle the first load next door to spoon out the humus around the plantings, returning twice more to leave Danute with her new garden beds chock-full of compost.

Carl, having finished his pre-party chores, wanders over with a beer in hand to investigate all the activity. After watching me tuck his grass clippings in with the leaf mold, he volunteers to retrieve the remote thermometer sensor he keeps in the dashboard cubby of his car. Carl owns a foam-installation business and uses the pistol-shaped device to measure radiant heat that escapes through walls and around windows.

"You want some heat," advises Stu Campbell in *Let It Rot!* But he cautions: "Heating can be tricky if it gets out of control. Earthworms are killed at 130° F, and they will not stick around and endanger themselves for very long in temperatures that even approach

that figure. Azobacteria, the precious microorganisms that transform nitrogen gas into a form that plants can use, are killed at temperatures above 160° F. Excessive heat is far more dangerous than no heat at all."

I dig out a small bore hole in the front face of my pile, and Carl clocks it with his temperature gun. The digital readings flick onto the screen: 119° F... 123° F... 114° F... 124° F. I dig a little deeper and 128° F registers on the little LED screen. We make sport of aiming the probe into the maw, which I expand a bit with each round. The LED display tops out at 131° F, and I imagine we could have easily found hotter spots within. A temperature of 131 degrees (55° C) is the accepted heat treatment standard for pathogen inactivation, provided all particles are maintained at or above that level for at least three days, notes James McSweeney, also the founder of the aptly named 131° School of Composting. I fear for the safety of the earthworms but take some measure of pride in having created a hothouse of a heap. I think even Carl is impressed.

In summer, it
burns like a stove.
It can—almost—hurt you.
I hold my hand inside the heap and count
one, two, three,
four.
I cannot hold it there.
Give it to me, the heat insists. It's mine.
I yank it back and wipe it
on my jeans,
as if I'd really heard the words.
— Andrew Hudgins, "Compost: An Ode"

The heat is the exhaust of the countless microorganisms thriving inside what Hudgins likens to a "slow damp furnace" where "life must do a million things at once." At this point in its death spiral, my pile must surely rank among the liveliest places on earth.

"Every compost pile is a complex ecosystem of decomposition experts," writes Julia Rymut, creator of compostheaven.com. "The main groups of microorganisms in soil are bacteria, fungi, protozoa and actinomycetes. These tiny little creatures are major players in decomposition. In a teaspoon of compost, you may find up to 1 billion bacteria, 440-900 feet of fungal hyphae, and 10,000 to 50,000 protozoa."

I don't know how many teaspoons of compost are in my pile, but I imagine enough to add many zeroes to Julia's numbers. Bacteria are not only the most abundant microorganism in soil, accounting for up to 90 percent of all microorganisms in compost, they also produce enzymes that further break down the complex carbohydrates—the cellulose and lignin of woody plant matter—that are slowest to rot.

Keith Reid, author of *Improving Your Soil*, explains what happens next with this complex cast of characters: "Through their sheer numbers, these organisms are able to access most of the easily digested materials in the soil and incorporate them into their bodies, with the sole purpose of making more bacteria. In the process, they release nitrogen, phosphorus, sulfur and other nutrients that have been bound up in the organic matter.

"Bacterial and fungal growth attracts a whole population of tiny animals that feed on them in much the same way that cattle or

sheep graze a pasture," Reid adds. "These include protozoa, such as the amoeba and paramecium. Large numbers of mites and nematodes also fill this role. Not all of the nutrients consumed by the grazers are used for their own growth, and the waste they release hastens the cycling of nutrients into a form that plants can use."

Though the worms may have fled to the cooler outer fringes (as I was rolling a log away from the right-side wall, a strip of bark peeled off, exposing a teeming colony of red worms), my pile is a hot mess, a riot of tiny creatures, all living, loving, laboring, and dying.

Carl gets called back home, and, in short order, my pile is reconstituted. The right side, newly infused with a rich mixture of grass clippings and seaweed, rises tall. I groom the top with the rake so the heap is roughly symmetrical, sides and slope matching. Soon I'll trade in the pitchfork for a shovel, and my pile will be no more.

Acid Test

Today, a Saturday in the middle of July, is bright and clear after a week of hot, muggy weather that was finally broken by a thunderstorm. The rains gave the yard and gardens and my pile a soaking down through the root zone and then back up until the low parts of the lawn were squishy under my shoes. I threw a tennis ball for Miller this morning, and his paws unleashed sprays of water; the fuzzy ball spun out its own spiral galaxy of droplets.

Before I take the hooch bucket outside to dispense with the weekly food scraps, I open the fridge, aiming to make one last sweep through the vegetable bin. Tucked away in the back is a box filled with a dozen or so donut holes, leftovers from a local fundraiser the weekend before that I'd agreed to take home and promptly forgot about. The donut holes are now rock hard, so I add them to my pile, figuring that if it's worth any endorsement dollars, I'd be happy to say my pile runs on Dunkin'. I'd also like to copyright "Hummus for Humus," on behalf of all those forgotten containers languishing in fridges across the land.

Along the top tray on the inside of the door is a collection of vitamins and supplements. There's a bottle of multivitamins and two of fish oil supplements from a new year's resolution. Both the resolution and the bottles are way past their expiration date.

I debate whether they should go in the hooch bucket or the trash can and check online for advice. While I know certain prescription drugs are now creating all kinds of hazards across the food chain when flushed down the toilet or sent to a landfill, I can find no authoritative counsel on whether multivitamins pose the same risk.

I find some concern, especially about small children coming into contact with excessive amounts of iron, but after reading on one garden forum the comment, "hey, the pills contain minerals that are naturally occurring anyway," I decide to chuck them into the hooch bucket. The fish oil capsules will surely wash safely away, and I figure a handful of one-a-day vitamins ingested by a pile the size of mine, soon to be spread across a third-acre of grass and garden, is an acceptable way of recycling such nutrients.

I set the hooch bucket next to a plastic bag of shredded paper from the office. What began several months ago around tax time as an impromptu gesture has now been added to my job description. Whenever the small bin of the finance department's paper shredder fills up, "the compost guy" gets the call to empty it.

It's a task I am happy to have taken on. For one, the shredded strips of crinkled white paper mix well with the fresh grass clippings and watery kitchen slop I add through the year. For another, I see my recycling at the office as a fairly easy way to practice what I preach at the place where I spend most of my waking hours.

And there's this: I was a working journalist for more than twenty-five years and until recently an avid consumer of printed magazines

and newspapers. Though I've switched to an online subscription, for decades I spent the first hour of every day reading *The New York Times* through and through. I read once, perhaps not in the *Times*, that it takes one tree grown for pulp to produce five fat Sunday papers. Over my career consuming or creating the news, I figure I'm at least partially responsible for the destruction of a sizable forest worth of paper. To recycle even a tiny portion of that highly processed product in my backyard is a way to atone for being an ink-stained wretch for so long.

Here's what the Cornell School of Horticulture has to say about it: "Several paper products—especially newspaper and cardboard—are useful in the garden and landscape. While it provides no nutrients, paper is organic material, made primarily of wood fibers. It decomposes slowly but provides structure when used in a compost pile. Shredded newspapers are good paper choices for composting or digging into soil directly. They decompose well when mixed with high nitrogen products such as manure."

Decades ago, as a young whippersnapper living in New York City, I helped a buddy fill up his parents' Volvo with a half-year's supply of the *Times* from their co-op in lower Manhattan. Our orders, in exchange for spending a long weekend at his parents' summer cottage on the North Fork of Long Island, were to spread the newspapers wholesale across a part of their garden. We carpeted the weedy ground with overlapping sections laid tight together, shingle style, tightening the weave around any perennials. In another part of the garden, the previous year's "all the news that's fit to print" had smothered the growth of weeds as it slowly degraded from newsprint to papier-mâché.

I grant the Cornell horticultural experts their concerns about adding shredded office paper to a compost pile. But I have to trust the roly-polys and their microbial allies that inhabit my pile to digest trace amounts of toxins and render them more harmless.

One piece of paper even the pillbugs have no taste for? Those little stickers on supermarket fruits and vegetables. In "Even Composting Comes with Sticker Shock," a Food column in *The New York Times*, Kim Severson writes: "The most bedeviling problem for the company that turns most of the Puget Sound region's kitchen waste into compost is on a piece of fruit. Almost every piece of fruit.

"It's that little sticker that tells you whether the fruit, and many kinds of vegetables, are organic, where they came from and which code a supermarket cashier should punch into the cash register.

"At Cedar Grove Composting, which every year turns about 115,000 tons of food and other waste collected from restaurants and home kitchens into dark compost for both gardens and larger construction projects, those stickers are a huge headache. In what seems something of a fool's errand, the company distributes sheets that look like Bingo cards. Fill one with twenty fruit stickers, and you get a free bag of compost. So far, they've given away about 800 bags."

I always try to peel off the stickers when I'm cleaning or prepping store-bought food, then tease the annoying things off my thumb with a flick of the finger into the trash can. Still, I'm amazed by how many crop up when I'm turning my pile. I wish they could make those stickers as perishable as the fruits and veggies they come on.

At this point in summer, my pile is easy to work with and through. I dig a trench along the backside that becomes a hole that allows me, for the first time since November, to reach the core of my pile, the very bottom of the very center. Using the straight-tined pitchfork, I tease out some surprisingly whole leaves and turn them onto the sides, where they mix with mats of rotting grass. I'm in virgin territory. In go the kitchen scraps and the shredded paper, mixed with the crusty leaves scraped from the sides and along the bottom edges of my pile.

With the pills and the paper tossed into the dark, damp middle of the heap, I get what I consider a bright idea—I'll gather up the mound of maple seeds rotting away in the other back corner of my yard and bury them as well. Since I swept them off the driveway and along the street, the mass of maple whirligigs has been left to molder away next to the brush pile. Lately, I've been kicking myself for not having blown more of the winged seeds off the wood-chip mulch that rings the perimeter of my yard. Over the past month or so, hundreds have sprouted, and I'll either have to pluck them by hand or smother them with a new supply of chips this fall. Dispatching the moldy maple seeds en masse gives me a small measure of payback. Forever grounded, the winglets will contribute rich stores of natal nutrients.

Here's what continually amazes me about my pile: Over the past eight months I've added to the original mass of autumn leaves a score or more tubs of seaweed and salt marsh hay, hundreds of pounds of soggy kitchen scraps, a gross of eggshells, a bushel of coffee grounds, more piss than I'm willing to fess up to, reams of shredded paper, a surfeit of droppings from the back end of both

horse and hare, and, lately, enough grass clippings and pulled weeds to fill the bed of a pickup truck. And after everything I've thrown at it and into it, my pile still just looks like a heap of old rotting leaves, only now in crumbly bits.

Despite all my reading, exactly what has occurred deep inside this dark, dank compost heap—or more to the point, exactly what I've now got on my hands—remains an enigma. So before wrapping up yard work for the day, I take three Ziploc bags and label each with a Sharpie: #1 is "compost pile," #2 is "wood chip mulched perennial bed," and #3 is "vegetable garden."

The precise nature of my pile will soon be put to the test by the Soil Nutrient Analysis Laboratory at the University of Connecticut. The sample, as specified on the UConn website, calls for one cup, "taken from 10 or more random, evenly distributed spots in your sample area."

Though by now I figure my pile is the very definition of random, still I try my best to follow the instructions and extract representative tablespoons with the tip of a trowel from ten or so spots. I take care to flick away any identifiable bits of leaves and toss back a seashell or two.

The other two bags, my idea of control samples, are easier to get. One is filled with trowel bits gathered from a few spots along my perennial beds covered by mostly decomposed wood chips, the other with small scoops from the raised, fenced-in vegetable garden.

"Soil testing is an inexpensive yet valuable tool for assessing the fertility of lawn and garden areas. Test results indicate the soil's pH level, the amounts of available plant nutrients and the existence of nutrient imbalances, excesses or deficiencies. Soil testing eliminates the guesswork many gardeners face when deciding the kinds and amounts of fertilizers or soil amendments they should purchase and apply... The standard nutrient analysis will provide the soil sample's pH, the available amounts of phosphorus, potassium, calcium and magnesium, extractable micronutrient levels and a lead scan... Separate analyses offered by the lab include percent organic matter, particle size analysis (the relative amounts of sand, silt and clay), micronutrients, soilless media and soluble salts." – UConn College of Agriculture, Health and Natural Resources

I seal the Ziplocs and stuff the three samples into a small box with a check and a form, marking the option that requests the sample be tested for organic matter and adding a note that I've included the mulched garden bed sample because I'm concerned that the rotting wood chips may be sucking nitrogen out of the soil. In all, the test costs $10 a sample, and results are said to be prepared in seven to ten days. So it's come to this: I'm digging up dirt on my own garden.

Off the Charts

I spend an anxious week awaiting the results of the soil test samples sent off to the UConn lab.

Not that there's anything I can do about it. I know all that's gone into my pile and of its seemingly infinite capacity to absorb and transmute the raw materials into a finished product that is not fully understood by scientists but invariably useful and productive to gardeners and farmers of every kind. It brings to mind this grace note from Wendell Berry, upon coming across an old bucket hanging from a fence filled with leaves:

"Rain and snow have fallen into it, and the fallen leaves have held the moisture and so have rotted. Nuts have fallen into it, or been carried into it by squirrels; mice and squirrels have eaten the meat of the nuts and left the shells; they and other animals have left their droppings; insects have flown into the bucket and died and decayed; birds have scratched it and left their droppings or perhaps a feather or two. This slow work of growth and death, gravity and decay, which is the chief work of the world, has by now produced in the bottom of the bucket several inches of black humus. I look into that bucket with fascination because I am a farmer of sorts and an artist of sorts, and I recognize there an artistry and a farming far superior to mine, or to that of any human." – Wendell Berry

I wake up early on a hot, dry Saturday morning in late July. Later, when my sleepyhead son finally wakes, we'll trek into New York City with our bikes to spend the afternoon exploring the city. Old men don't need their beauty sleep, so I set to some gardening chores before hauling out the bikes.

One way to think of my backyard is vertically, perhaps a notion on my mind as I look forward to cycling through the canyons of downtown Manhattan. Seen this way, there are four separate layers to my landscape: the ground, which includes the lawn, groundcover, and mulch beds; the annual and perennial bushes, flowers, and vegetables that grow chest high each summer; the larger shrubs and small trees, like the dogwoods, lilac, redbud, crabapple, and the collection of young hardwoods I'm raising; and the mature canopy trees—sycamore, pine, and maple among them—scraping the sky.

My property is positively stacked. Each season all those trunks, limbs, and branches grow taller and wider, which calls for some measure of cultivation. I pass by my pile to get the extension clipper and saw from the shed. The flexible, two-part pole is long enough to reach the sucker limbs of the crabapple that grows, almost unseen,

along the fence between the Rosens' house and a blue spruce I planted years ago. Next to it is a patch of fifteen-foot-tall privets rising just beyond the fence that divides our properties.

An introduced species of a semi-evergreen shrub found throughout Asia, privet has long been an American landscape fixture, defining Southampton estate and suburban backyard alike. Early each summer the privet's fast-growing limbs are hung with white flowers, beloved by bees, which in turn produce heavy boughs of purple berries that the early-arriving robins gorge on while they're waiting for the yard to thaw and give rise to their earthworm prey. The berries the birds don't eat produce many seedlings that sprout in the wood chips on my side of the fence. I admire how privet has set down such deep domestic roots, but the gangly limbs now crowd out airspace in my garden. I give them a good whack.

Next, I prune a sycamore that sprouted in the pachysandra bed along the west side of the house. I know the admonishment to never let a tree grow next to a house, but I am an indulgent gardener, and I've been amazed to see how fast the sycamore has grown in just a handful of years. It now rises ten feet above my second-floor attic and spreads wide enough to cast the entire west side of the house in afternoon shade. The benefit of a cooler home offsets any concern I have about the tree's roots causing problems with my foundation. My house sits on a cinderblock foundation and has a dirt floor under the crawl space. If a root wants any part of that creepy-crawly space, have at it.

To keep the sycamore from scraping against the side of the house and roof, I trim it espalier-style. Though now only half a tree in some respects, it's handsome and robust. I'll let it go another year or two

and then decide its fate. The tree this intrepid young sycamore hails from lords over the front corner of my yard in its majestic if messy way. It is the largest living thing in the yard, if not neighborhood, and I spend more time picking up after it than all the other trees on my property—in addition to dropping its fluff-filled seedballs across the yard, it sheds leaves pretty much throughout summer, and, after the first heat wave each year, its bark peels away from the trunk like skin with a bad sunburn. Whole chunks flake off from top to bottom, littering the lawn beneath the tree.

I read in *Around the World in 80 Trees*, a book Cole thoughtfully gave me for Father's Day, that this habit of bark shedding makes the trees especially well-suited to urban life. Author Jonathan Drori explains that the sycamore has a special trick that helps it thrive in polluted air. "Its bark is brittle and, because it cannot adapt to the growth of the trunk and branches underneath, it drops it off in flakes the size of a baby's hand. The bark is dotted with tiny pores, a millimeter or two across, called lenticels, which allow the exchange of gases. The ability of the tree to slough off a layer of grime that it has removed from the atmosphere helps to keep both this city-dweller and its human companions healthy." Usually, I simply mow the thin, curved sheaves of dropped bark into mulch, but during the dog days of August there's enough to rake and pile onto my plastic tarp, which I drag catty-corner across the yard to deposit in the brush heap.

My pile sits untended, but if I thought today was a day away from compost, boy, was I wrong!

Wheeling our bikes off the Metro-North train and up and out of Grand Central, Cole and I cycle down Forty-Second Street to the West

Side bikeway to head south along the Hudson. I lived in Manhattan for a decade at the start of my career and cycled throughout the city, often along empty streets late at night, to unwind after the close of the weekly issue of the business magazine I worked for. Our destination is Governor's Island, where I'd read that a Civil War re-enactment will take place. I figure this bit of living history will stand in for some needed summer enrichment. Plus, I get a kick out of taking the free ferry across New York Harbor to the obscure old military installation, now furloughed.

Of course, Cole isn't buying even this hint of "homework" on a summer Saturday and makes his own detours. Coursing through the canyons of Wall Street, filled on a weekend with tourists instead of suits, we finally make it ashore just as the Civil War cannons and muskets fire their last salvos.

As we pedal to the southern side of the small island, we come across a section of old barracks that have found new life as Earth Matter, a hippieish, communal operation dedicated to... compost. I'm delighted to stumble across it, and my son is thrilled by the sight of chickens free-ranging about. We park our bikes and enter the fenced-in compound, past hand-painted signs that announce "Free Compost!"

As we wander about the rustic, barnyard-like facility, almost literally in the shadow of the world's foremost concrete jungle, we learn more. Earth Matter, founded in 2009, is a nonprofit organization dedicated to advancing the art, science, and application of composting in and around New York City. Its mission is to address the dual problems of resource recovery and healthy soils with a single solution: promoting the local composting of organic waste into a healthy soil amendment.

We walk past long windrows of compost, each one planted with a sign that gives its date of creation. I envy the small front-end loader parked beside the nearest pile. A demonstration area features a row of different composting setups and contraptions, from tumblers to worm bins to a variety of fenced-in enclosures. Call it a dis-assembly line. At a pile set up next to a screen made of small-gauge wire, a young volunteer with collegiate beard scruff offers a shovel and a paper bag of the kind you see used for coffee, inviting us to "sift your own compost, and take home a bag."

I take him up on the offer, and while I scoop a couple shovelfuls of what looks like dried wood mulch, he gives me his spiel. "Did you know that compost heaps heat up to 1,500 degrees as they cure?"

I set the shovel down. "Fifteen hundred degrees? You sure? That's pretty hot—like melting steel hot..."

"Fifteen hundred," he repeats.

"Not more like 150 degrees?" I counter.

"Nope. 1,500," he says with certainty.

I package my few ounces of kiln-fired compost and thank the young man. I admire his conviction, if not his facts.

Arriving back home from the city, I check the mailbox and find a business envelope with a return address of University of Connecticut. Inside is a sheaf of single pages, results of each of my three soil samples. At the bottom of the first page, in pencil, is a handwritten note.

Hi,
We did run your 3 samples as requested. Our standard nutrient test is not meant for compost, as the nutrients are all above our mineral soil limits on the analytical equipment...

I check the tabular results for the compost sample. On the left of the page are four rows labeled Calcium, Magnesium, Phosphorus, and Potassium. Stretching across the page are three columns, titled "Below Optimum," "Optimum," and "Above Optimum." The colored bars for all four key minerals extend into the last column, with Phosphorus stretching the farthest.

My pile is off the charts!

Another page is labeled "Modified Morgan Extractable," and on it I see that my pile has 38.2 percent organic matter—which makes me wonder what the other 61.8 percent consists of.

Time for me to dig a little deeper. Turns out, garden-variety soil is 50 percent water and air and roughly 45 percent minerals such as sand, silt, and clay. According to U.S. Department of Agriculture criteria cited by UConn, my samples classify as "sandy loam," mostly sand, with nearly a third silt and a smidgeon—about 4 percent—of clay.

Your typical backyard garden has 4 to 6 percent organic matter, the USDA tells me, while pointing out that the nation's agricultural land now has much less, as little as 1 or 2 percent. My vegetable garden clocks in at 12.2 percent organic matter, while the mulched bed of the perennial garden registers fully 20.1 percent organic matter. Healthy levels of organic matter in the soil provide numerous, essential benefits, which the soil scientists at USDA break down into three categories:

- Physical: Organic matter enhances soil stability, improves water infiltration and aeration, reduces surface crusting and runoff, raises water holding capacity, and makes soil easier to till and to seed, particularly "sticky" clay soils;

- Chemical: Increases the soil's ability to hold on to and supply essential nutrients, helps soil buffer or resist swings in pH change, and speeds decomposition of soil minerals, making the nutrients in the minerals available for plant uptake;

- Biological: Provides food for the living organisms in the soil, boosts soil microbial biodiversity and activity, which helps suppress diseases and pests, and enhances pore space or soil "tilth," which further improves water infiltration, reduces runoff, and makes it easier for roots to thread their way through the soil.

And therein lies the problem for suburban gardeners and, on a grander scale, Big Ag. For more than a century now, modern agriculture has relied on mechanized tillage to break up the soil for planting and poured on an unending stream of synthetic fertilizers to boost growth and herbicides and pesticides to control pests and weeds. These unsustainable practices are now baked into the system.

"Since colonial times, U.S. soil has lost about half its organic matter," says David Montgomery, a professor of geology at the University of Washington, in a Politico.com article titled, "Can American Soil Be Brought Back to Life?" He blames the invention of the plow, which made farmland amazingly productive by disrupting the virgin soil. But tilling, it turns out, kills off many of the microorganisms that build the soil. It churns up their habitat and exposes them to air; it

also makes it easier for soil to be washed off the land by rain and wind.

As President Franklin D. Roosevelt put it, when confronting the manmade crisis that was the Dust Bowl, "The nation that destroys its soil destroys itself."

To counter the increasingly infertile soil that remains, farmers have relied on synthetic fertilizer, primarily nitrogen, in ever greater and more costly amounts. In 1964, farmers were applying on average fifty-eight pounds of nitrogen per acre of corn. These days, the amount is closer to 140 pounds per acre. A 2022 report published by the AGU scientific society estimates the erosion in the Midwest is occurring at double the rate that the USDA says is sustainable. Since converting 99.9 percent of the original tallgrass prairie into farmland, 57 billion tons of topsoil have washed or blown away. God only knows how much of that nutrient-laden soil has flowed down the Mississippi and its tributaries, but enough to create a dead zone of hypoxic waters off the river delta late each summer that rivals New Jersey in size.

This "... but that's how we've always done it" approach to farming (and gardening) doesn't hold water, but it does hold sway. Thankfully, more and more gardeners are going organic, and composting, and some farmers are bucking tradition by adopting regenerative practices like no-till cultivation, planting cover crops, crop rotation, applying nonchemical herbicides and pesticides, and the more widespread use of compost on their fields. The *Politico* article cites an Indiana farmer who switched to no-till, using cover crops and adding cow manure on 4,000 acres, and since has seen his soil go

from 1.8 percent organic matter to 4.8 percent, and still ticking up. There's hope for us yet.

On the back page of test results, the handwritten note continues: *As to your mulch question—I wouldn't worry about it. The benefits of mulch far outweigh any problems with nitrogen deficiencies. Wood mulches, in particular, decompose so slowly that nitrogen deficiencies would not typically be seen here. If you added a finer, more rapidly decomposing source of high carbon organic matter, like sawdust, you could see nitrogen deficiency around quick growing annual plants, like peppers and zinnias. I use shredded bark mulches in all my perennial beds and usually fertilize once in May. I have never seen any nitrogen deficiencies. Hope these tests answer some of your questions. Dawn P*

I wonder if I've given the vegetable garden too much of a good thing this summer by tucking in so much freshly minted compost. "If organic matter is above optimum, none should be applied for a year or two to allow the level to decline through normal breakdown," suggests the UMaine Cooperative Extension.

The Nitrate-Nitrogen level for my compost pile is 150.1 parts per million. Nitrate (NO_3) is the most important source of nitrogen available for plants; this test measures the amount of nitrogen in the soil that can be absorbed immediately. Uncultivated land has a nitrate level of about 5 parts per million; 19 PPM is in the range for fertilized gardens or cropland, which is right where the samples from the mulched beds and garden land. A nitrate level of 150 PPM explains why humus is used as a soil amendment, not a substitution for it.

A final page in the handout, "Interpretation of Soil Test Results," explains another mystery: What the heck is pH? "Soil pH is a measurement of a soil's acidity, which affects the availability of necessary plant nutrients." The pH scale ranges from 1 to 14, with a pH of 7 being neutral. Values below 7 are considered acidic while those above indicate alkaline conditions. My pH readings are 7.0 for the compost sample, 6.6 for the mulched flower bed, 5.5 for the vegetable garden.

"Most garden plants prefer a pH between 6.0 and 6.8. Notable exceptions include acid-loving blueberries and ericaceous plants like rhododendrons, azaleas and mountain laurel. These plants prefer a pH of 4.5 to 5.3. The majority of Connecticut soils tend to be acidic with pH values ranging from 4.8 to 5.5 due to the geology and climate of the region."

Overall, I'm pleased with, if not downright proud of, the test results. My pile turns out to be rich in nitrate and other nutrients. Though heaped with organic matter, the vegetable garden, with its base of native soil, remains true to its slightly acidic nature, which I can address by adding some lime or more seashells from the seashore this fall. And I'm relieved that years of covering my perennial beds with wood chips and chopped leaves haven't turned it too acidic or nitrogen poor. I have a science-based answer to why my flowers thrive so well—and perhaps why the rhododendrons haven't. I'm also happy to have received such personalized attention from a government bureaucracy. Thanks, Dawn!

AUGUST

A Day in the Moss

The saying "a rolling stone gathers no moss" is attributed to Publilius Syrus, a Syrian slave who became a popular performer in Caesar's Roman Circus. He's the wag who also gave us such maxims as "honor among thieves," "it may not be right but if it pays think it so," and "no man is a hero to his valet." Imagine how popular he'd be today if he could trade in his papyrus for a podcast.

Aside from sparking a soulful tune that plays on loop in my head as I dilly-dally about the garden, Publilius's adage brings to mind a fascination I have with the mosses and lichens that populate the shadier precincts of the backyard. Some years ago, I came across a curious way to grow lichen and moss—by slathering a rock with spoiled yogurt. Having just such a container in the back of the fridge, I took it out to a rock I had unearthed from the lawn and rolled over to an empty patch of mulched wood chips and pine duff in the southeast corner of the yard. The sizable stone, near boulder in size and worn smooth and round by glacial drift, was fetching in shape but raw in appearance. No telling how long it had been buried.

Trust me, you haven't fully become a garden obsessive until you give a yogurt massage to a rock. But sure enough, years later the

rock is resplendent, cloaked in a mantle of green moss and blue-gray etchings of lichen. I prize the look of this stone, so much that I hardly ever sit on it. A few weeks ago, I noticed it had begun to subside into the deep layer of pine needles and wood chips, so I rolled the stone over and propped it back up with a brace of smaller stones plucked from the yard. I suppose if you put a rock on a pedestal, it becomes a statue.

If we think about lichen at all, most of us know them as curious splotches on old stones, weathered tree trunks, and, fair to say, gravestones—subtle signs of immutable decay. I see lichen as enduring decorations from deep time. The small boulder sitting in the corner of my yard sports more than a pretty patina; on its mottled surface is a gathering of living creatures, neither plant nor animal, thought to be among the earliest land-dwelling forms of life. Or so I find from garden writer Margaret Roach in a *New York Times* article that sheds light on an inscrutable species found on every continent, covering an estimated 8 percent of the planet's land.

Quoting two self-described lichen lovers and authors with ties to the New York Botanical Garden, Roach describes lichen as "composite symbiotic organisms, an intensive cooperation between a fungus and an alga or a cyanobacterium, and sometimes all three. Most of a lichen's structure is the fungus. The alga lives with it, and in return for shelter it provides photosynthesis, producing sugars that sustain the fungus. But the two are not alone.

"'In many ways, lichen are miniature universes,' Dr. Allen and Dr. Lendemer write." Lichens can live on anything. All manner of animals nibble away on them for sustenance, from mice to moose, and the many lichen species are avid, if leisurely, decomposers.

"Take note, gardeners," adds Roach: "Lichens help with soil formation by accelerating the breakdown of rocks. They perform nutrient-cycling, too, as the cyanobacteria in them fix nitrogen from the atmosphere, converting it into a more usable form.

"And when lichen falls on the soil and breaks down, Dr. Allen said, 'it's a little packet of fertilizer—a fertilizer application.'" I'm also intrigued to know the quip that gave The Temptations a No. 1 hit and launched epic careers for Mick Jagger and Keith Richards has been adapted to suit the ever-evolving ears and interests of a given time. The 1825 *Dictionary of Scots Language* homes in on the phrase, explaining that "Any gentleman, whether possessing property or not, who was popular, and ready to assist the poor in their difficulties, might expect a day in the moss, as they were wont to term it."

"'A day in the moss' referred to cutting peat in bogs, hard work in preparation for winter," Wiki explains. "An itinerant 'rolling stone' will not likely feel the timely need to 'gather moss', by applying for access to a community's peat bog." Gathering moss is just what I'm up to after receiving a notice from the company that holds my home insurance policy. Evidently, they do drive-by inspections of their policyholders' properties, and they dinged me for having an excessive amount of moss on my roof.

I borrow an extension ladder from the collection that Michel keeps stacked in his backyard and align it to the slope of the north-facing roof in my front yard, first spreading a canvas drop cloth under the overhang. Employing a long-handled, straight-edged garden hoe—a tag-sale find seldom used for its intended purpose but very handy for this one—I scrape clumps of moss and patches of lichen that have flourished in the edgings of the crumbly old shingles. Before long I've cleared the shingles from roofline to dripline, at least to my satisfaction, and gleaned a smattering of unmoored moss and flecks of lichen on the drop cloth.

The most whole pieces of moss I place on a few of the old logs I've parked over the years among the perennial beds to serve as "rotting log hotels" for bugs and pollinators and other overwintering critters. Several come from the pile's containing walls, too far gone to support my weight but still of use in the decomposing construct that is my backyard.

"'Biodiversity' as a word sounds rather dull and a bit abstract. Played out on the ground it is something else: the difference between the numbered title of a symphony and its glorious complexity unwrapped in a concert hall. Every rotting log is a small world. The underside of a leaf is a realm to a greenfly; a crack in the bark of a beech tree is a capacious and secret hideout. They all fit together in a jigsaw that remakes its own pieces month by month. Rot is creation in the underworld." – Richard Fortey

I unfurl the rest of the scrapings across my pile to mix in with the near finished compost soon to be spread across the lawn and garden. A rolling stone may gather no moss, but the stuff in my yard sure does get around.

Nip and Tuck

It's a bright, breezy Saturday morning in late August that brings with it the first hint of fall. There's a crispness to the air, and looking out across the backyard toward my pile, I see a single maple leaf, tinged with gold, flutter to the ground.

That's all I need to get started with one of the most pleasurable annual rituals of backyard gardening—dividing and transplanting perennials. Though most people think of planting as a spring thing, I favor early fall. I've seen what's grown and flourished, what works well here or there and what doesn't. Fall planting allows plenty of time for the roots to reestablish themselves and grow strong enough to survive a long, cold winter. Plus, with the compost pile now simmering to completion, I have time to fiddle or to correct any garden miscues I've made along the way.

> "I think having land and not ruining it is the most beautiful art that anybody could ever want." – Andy Warhol

This year, I've had my eye on a patch of sensitive ferns that has spread across the perennial bed on the western side of the yard. I moved a few sprigs there from elsewhere in the garden several years ago—mostly to insert some summer greenery that doesn't appeal to the deer. But I've since trimmed back the overhanging privet bushes and the crabapple tree along the fence to provide more sunlight for a redbud tree I planted in the middle of the patch. Now sunburned, they distract from the pretty little redbud, which shows promise of becoming a fine specimen of a tree for this side of the yard. I've spotted several shadier, less crowded spots in the garden for some of these truly sensitive ferns to go.

Just down the way from the fern patch are two hydrangeas. Over the years, the plants have grown large and fused into one, crowding each other and the clearance-sale pin oak I planted behind them. I decide to move the hydrangea that's closest to the oak to the sunny spot vacated by the ferns. It's the outdoor equivalent of moving the furniture around. It costs nothing and gives the garden a new look.

I head over to my pile, which sits like a Sphinx in patient slumber, its weathered flanks covered with crumbly detritus. I've already harvested several wheelbarrows from one side for Danute's new kitchen garden and will withdraw that much or more for the tending and transplanting of perennials. I've been a good steward of this heap for the better part of a year. I've lavished it with offerings, mulled over it, and tossed and turned every bit of it. I will miss this particular edition of my pile, but it's time to return its makings whence they came. I use the pitchfork to fill the wheelbarrow, not bothering to sift the clumps.

I know some garden experts now advise not to use any soil amendments when transplanting but it's hard to disagree with Elsa Bakalar, author of *A Garden of One's Own*: "Never plant without a bucket of compost at your side." For a backyard composter, nothing compares to digging a hole in a garden bed, adding a tender plant plucked from earth or pot, then tucking it in with heapings of fresh compost. For a new planting, compost is both an insurance policy and a deposit guaranteed to pay dividends. As long as I add water, virtually every transplant thrives. Compost is such a surefire potting-soil mix, I even have the confidence to begin the fall transplant season in late August. There's still the risk of a late-season heat wave, but with some extra watering, the plants will have plenty of time to properly root themselves before the frost and freeze of winter set in.

Freeing an already established plant, though, calls for a gentle hand. The roots of this particular fern spread out shallow along the ground. I circle around the edge of the root mat with the straight-tined pitchfork, loosening its grip so I can pull it up like a thick carpet. I toss the tangled pelt onto the compost that fills the wheelbarrow, separating the roots and stems into three clumps.

These will go up against a low rock wall that fronts a small shade garden along the main street, next to the Rosens' yard. Two Japanese maples, transplanted a decade ago from coffee cans given to me by my neighbor Pierre, bookend the raised bed, which is contained by a border of stout rocks wrested from under the yard. The trees are now fifteen feet tall and form a scarlet screen between yard and street. Between them is a dogwood I planted as an Arbor Day giveaway twiglet years ago and protected from the deer until it was

head high. I took the wraps off this past spring, and so far they've left it alone. A profusion of bleeding hearts fills the rest of the bed; their branches of dangly pink and white are another springtime show. The ferns will fit in perfectly.

The year before last, I'd thinned the bed and given a couple baby bleeding hearts to Joanna, who transplanted them on the shady side of her house, where they bloomed fetchingly this past spring. Her daughter, Alyssa, loved their look and asked her mom if I could give her some. Alyssa was a high school senior when we first moved in and was Cole's one and only babysitter in our home. She's since graduated college and become a teacher, married with a baby girl of her own and a new house in the town next door.

Though the fleshy stems and three-lobed leaves of this spring ephemeral from Siberia are withered in the summer heat, I take care to unearth three clumps, keeping their rhizomatous roots as intact as possible. They'll make a lovely, albeit DIY, housewarming gift for Alyssa and free up space for the native ferns, into whose place the hydrangea will go.

I plop the ferns into their newly made holes and surround their roots and stems with shovelfuls of compost, emptying the rest of the wheelbarrow around the slender trunks of the Japanese maples, the dogwood, and the remaining bleeding hearts. Plants don't do well growing in pure compost—it would be like breathing 100 percent oxygen—but a little trowel work, some foot pressing, and water is all it takes to thoroughly mix the new humus with old dirt.

The hydrangea blends right in with the rest of the plantings, here a mix of coneflowers, Joe Pye, and black-eyed Susans, with plenty

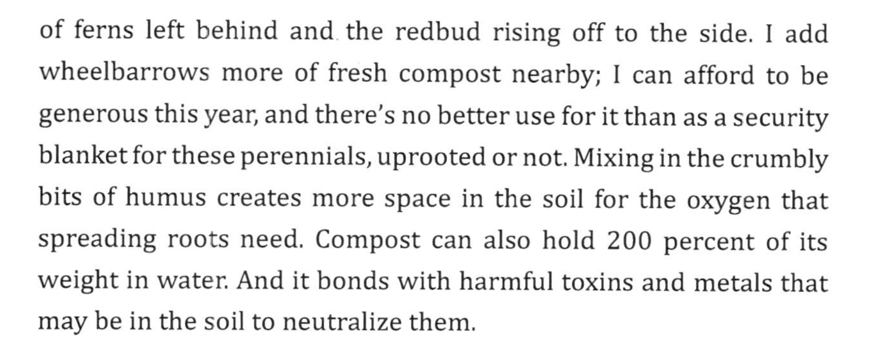

of ferns left behind and the redbud rising off to the side. I add wheelbarrows more of fresh compost nearby; I can afford to be generous this year, and there's no better use for it than as a security blanket for these perennials, uprooted or not. Mixing in the crumbly bits of humus creates more space in the soil for the oxygen that spreading roots need. Compost can also hold 200 percent of its weight in water. And it bonds with harmful toxins and metals that may be in the soil to neutralize them.

Though some types of hydrangea are native to America, I'm pretty sure these legacy plantings hail from Asia. I keep them as I do the antique rolltop oak desk bequeathed to me by my parents when they downsized to Florida, an heirloom that doesn't quite fit with the décor but is a part of the household nonetheless. Besides, both are too hard to get rid of. I also like the fact that hydrangeas allow me to tinker with the color of their blooms. With my slightly acidic garden beds, their natural color is purplish pink. But years ago I came across a bit of hand-me-down wisdom that says if you tuck a few rusty nails around their base, the blossoms will turn blue. Sure enough, it's true—the oxidizing nails turn the soil more acidic—and I always have a place to dispense with the nails I bend and loose screws I come across.

Finally, I have another chore: to find a repository in which to stockpile the continuing supply of kitchen trimmings and other compostables my pile cannot accommodate as it matures. There comes a time each season when I cut off the compost heap from having to ingest fresh fodder and instead create a new starter batch. It helps that each August, Danute takes her four girls back to her homeland, to visit family in Budapest. Some years, Michel joins

them for side trips to Switzerland or the Algarve. When he doesn't, he seldom contributes to the food waste bin Danute keeps for me outside the back door.

Ever the packrat, Michel has a collection of garbage cans stacked alongside the house, several of which he never uses. I borrow one with a lid and set it up beside the log wall that I've splayed to get to the side of my pile. I layer the bottom with sycamore bark gleaned from the brush pile, then cover those scraggly bits with half of my regular supply of shredded paper to lay a base of absorbent material. After topping this off with a pitchfork of not-ready-for-primetime compost, I upturn my kitchen hooch bucket, then cover that small batch of food waste and coffee grounds with another forkful of raw compost. In short order, the garbage can is near half full. After mixing up the mess with a pitchfork and adding a bit more compost and shredded paper, I put the lid on, flipping up the handles to seal it. The lid is tight enough to keep out flies, and the few stray worms I've likely tossed in will have the compostables to themselves for the next few weeks.

I plunge into the heap with the pitchfork, relishing the simple task of digging through the loose matrix of crumbly humus. I search for sulfurous spots of matted grass but find only stray flecks of yellow-green. My pile has burned through the infusions of grass clippings I've added to it on through summer. Digging further into the side, I come across a cache of musty maple seeds, the very ones I added to the backside of the heap several weeks ago. How ironic that, at long last, I've reached the very heart of my pile only to find what I never wanted to see again. I unseal the garbage can nearby and stuff it with all the zombie maple seeds I can find.

Garden Island

It's the final weekend before the Labor Day holiday. With my pile on deck for distribution and Cole heading off to college in a few days for freshman orientation, it feels like the end of an era. Now that my son has aged out of backyard play, and the mutt and I are ever less frisky, I can see the lawn losing ground to other priorities. I once aspired to create the sort of turfgrass that Katherine S. White described as "a soft mattress for a creeping baby." Now I scope out anywhere I can to trim its edges or let it grow wild as meadow in the middle.

"Reducing the area we currently allocate to lawns is a necessary and logical consequence of ecological landscape design," writes Douglas W. Tallamy in *Nature's Best Hope*, in which he advocates for converting backyards into an interconnected network, what he terms the Homegrown National Park.

"I don't think it's likely (nor do I suggest) that we will ever abandon the lawn as a landscaping tool. Turfgrass species are perfect for areas where we walk, for example, because they can withstand moderately heavy foot traffic.

> But transitioning from landscapes in which wall-to-wall turfgrass is the default, to landscapes that thoughtfully use lawn as pathways through savannahs of spreading native trees, native forbs, and warm-season wild grasses is now entirely within our grasp and presents a new way to demonstrate our creative abilities." – Douglas W. Tallamy

I've already plotted out my next turf-removal project. In the front yard, the young tulip poplar and companion red oak grow on the island of wood-chip mulch near where the late, lamented tulip magnolia once stood. Underneath and alongside them is a mix of ferns, lilies of the valley, and native wildflowers interspersed with a stand of lilacs, transplanted a few years back from sucker saplings that rose from the big old bush next to the back door. In the springtime, the oval-shaped patch is dotted with daffodils, as well as fast-spreading clumps of bluebells. There's even a yucca plant, sporting two big spikes of silvery green blades and white flowers. Across the lawn, about ten paces away, is a fine pink dogwood in its prime rising from a small circle of mulch of its own.

My idea is to connect the two by removing the grass between them, turning the two islands into more of a continent. Say, like Australia. It's an oblong stretch of turf about thirty feet long by fifteen feet wide, set three paces from the street, leaving a strip of grass along the road for the neighborhood dogs and their walkers. I rarely set foot on this part of the lawn unless it's to mow it.

In coming days and weeks, I'm eager to plant this new ground with perennial flowers from other garden beds around the yard that have grown too crowded and need space to thrive further. I'm also looking forward to filling in this sunny new patch with some choice purchased plantings, natives praised for what they bring to the pollinator party and for the wildlife they attract. Except deer, of course.

It will be my biggest landscaping project since establishing the garden beds along the perimeter of the property and the house itself after moving in. Good thing that Cole has promised me he'll pitch in before he heads off to college. Lately, he's begun to view the yard in a different light; it warmed my heart last weekend to see him give a girl he's courting a tour of the backyard. I hung back on a chair under the umbrella on the patio, but within earshot as he called out, "Dad, what's the name of this one?"

"Joe Pye weed."

"And this one?"

"Black-eyed Susan."

"Funny the names they give to flowers," said the girl. Then they scooted back inside, using the excuse of being attacked by butterflies. Ah, young love!

So it was these two lovebirds I suppose who sparked the idea of creating a new flower garden out front. I'm inspired not just by Tallamy's Homegrown National Park initiative but also by an unrelated but overlapping ecological movement: the Pollinator Pathways project.

Conceived in 2007 by Seattle-based designer Sarah Bergmann, Pollinator Pathways is a participatory ecology effort that encourages public and private properties to restore or create pesticide-free plant habitats for pollinators. The idea is to establish enough safe and nutritious places for bees, butterflies, hummingbirds, and other pollinating insects and wildlife to rest, eat, and breed. Grown close enough together (native bees have a range of about 800 yards) and near larger parks and preserves, Pollinator Pathways aims to "defragment" the urban/suburban environment so it will be able to support sustainable populations of wildlife. The local branch of this grass-roots initiative, the Westport Pollinator Pathway, is most active on this front, having forged partnerships with a host of garden clubs, nature centers, and eco-groups to create new pollinator-friendly spaces all around town, both on public ground and private property.

Tallamy has even more ambitious plans for Homegrown National Park. He wants to repurpose half of America's vast lawnscape for ecologically productive use. An entomologist by training, he argues that we can no longer rely on national parks and other "wild" places to sustain a viable ecosystem. Noting that much of the 40-plus million acres of cultivated grass in the U.S. is laced with herbicides and pesticides and about as useful to a pollinator as a parking lot, he says that if half of all that wasteful turf could be replaced by more pollinator-friendly gardens and greenscapes, a functioning network of habitat the size of ten Yellowstones could be created by untrained citizens at minimal expense and without costly changes to public infrastructure.

So that's how I came to be standing in the front yard, edger in hand, to rewild my front lawn and turn a good-sized chunk of it

into a new garden bed, adding plenty of flowers and curb appeal along the way. I use the edger, a curious tool with a blade shaped in a semi-circle (another vintage tag-sale find), to outline the perimeter of the new garden bed. I figure I'll simply dig up the sod with a spade, then turn the shovelfuls of grass upside down to decompose in situ. It's far too much sod for even my covetous old pile to absorb, and besides, there's a slight downward slope from the curb to what will be the center of this new garden bed, so I'll need to keep all the soil I can in place.

I've nearly finished etching the outer rim of new garden bed when sleepy-eyed Cole ambles outside. After shooing him back into the house to trade his flip-flops for sneakers at least, I hand him the spade and try not to over-instruct him on how to turn the sod over, root-side up. Just think of it like our old upside-down Christmas tree, I tell him. My hope is that by exposing the roots to the late-summer sun and burying the green growth, the grass will soon wither away. What scrapes of sod that do resurface can be dispatched to the pile. I'll cover the bare ground with this year's abundant supply of compost and, I expect, a fresh batch of wood chips later this fall.

Since I won't be planting right up to the border, I trade the edger for the flat shovel to remove a spit of turf (a "spit" is the length of a spade) all around the perimeter, flipping the patches of sod onto the sections of lawn already turned by Cole's spadework. Clean margins for the miraculous child his mother and I thought we'd never be able to have.

The earth is easy to till, and he and I make good progress toiling away in our separate lanes, Cole stopping only occasionally to

check his phone and me for a very refreshing morning beer. I'm celebrating the fact that, for once, I've got help doing a garden project that calls for a strong back.

For many people, certainly me, gardening is a meditative, almost solitary pastime. That's all well and good, as numerous studies have shown that the rewards of working productively in the garden include improved mental health. Psychologist Seth Gillihan highlighted two other attributes of gardening in a podcast with Joe Lamp'l, host of the PBS show "Growing a Greener World": "Few things boost our well-being like good relationships, and gardening offers ample opportunities to connect with others," said Gillihan. "Gardening provides a connection not just to other people but to our world."

"It's a collective effort," added Lamp'l, who is also the creator of joe gardener, an online guide to organic garden resources. "We're all better together when we share our experiences. We all have an innate connection to the earth."

Before too long, Cole's young friend stops by with an egg sandwich and an iced coffee for him. They head back inside, and for a time I fear I've lost him for the day, with more than half the new garden bed yet to turn. I take up the spade and plunge in. The rich topsoil reaches down through the root zone to denser clay further below, pockmarked with the ubiquitous little rocks. I'm relieved to spot few white grubs and plenty of earthworms.

Eventually Cole and his sweetheart reemerge. I hand him the spade and take my own break. Cole's friend says she's keen to pitch in but is wearing only sandals. I give her the half-moon edger, suggesting she could use it to break up the clumps Cole is turning over. That soon turns the work into a game, with Cole racing to keep ahead of

the girl's playful jabs. Her true value to this enterprise comes when my boy, flexing further, takes off his sweat-soaked shirt. I know then that we will finish turning this stretch of lawn into garden today.

The backyard stands as a ready-made source for perennials to transplant. The purple coneflowers and black-eyed Susans have flourished so well in the small fenced-in garden I keep next to the back patio that they're beginning to crowd out the vegetables. And after deciding the other day to undertake this impromptu project, I've already placed an order for some native plants from a local conservation group hosting a fall sale. I picked hyssop, viburnum, a winterberry bush, whose bright red berries provide food for birds deep into winter, and winged sumac, to backstop and eventually replace the lilac. The organization, Aspetuck Land Trust, has its own Green Corridor Partners initiative, which invites area gardeners to plant native, switch to organic or zero-emissions lawn care services, and stop using pesticides. Between this community effort and the others, our new garden will certainly be on the pollinator map.

This garden won't be strictly native; according to Tallamy and others, in terms of enhancing the biodiversity of native insects and animals, the benchmark to shoot for is having no more than 30 percent of plantings be exotics. I will continue to savor the lilac scents of spring, as will the dog walkers passing by; the doves can enjoy their feasts of cleome seeds. I suppose I'm accepting of these established immigrants because so many of my neighbors are from elsewhere and they have put down roots here just fine, thank you very much.

As Arthur Shapiro, a professor of evolution and ecology at the University of California at Davis, explained in a 2022 *Washington Post* article about how gardeners are planting natives to cope with climate change and biodiversity loss, "It is silly to try to re-create

conditions obtained in the past, when the boundary conditions have changed irrevocably. It's not restoration. It's a gardening project. We should be studying the 'novel ecosystems' that are arising spontaneously under our noses to see what we can learn for conservation and management. It's the history of the world: Change is the normal condition. Stasis is abnormal."

To plot things out, I had drawn up a rough schematic of the new garden on a piece of paper the other night, a map with X's marking the spots where I intend to plant and with what. I left the sketch on the coffee table and, when I returned to it the next day, saw that Cole had scribbled across the top, "Nice!" Another milestone reached: My son is now grading his old man's homework.

SEPTEMBER

Labor Days

I step onto the porch with a Saturday morning cup of coffee in hand. I'm back from dropping Cole off at college. I was happy with his choice of school, a small liberal arts college in western Massachusetts that's a scenic three-hour drive due north. On the way home, I spotted several small dairy farms that I bet will allow me to fill a bucket with manure, so I'm sure I'll come up for visits.

But now I'm just a cat and dog dad, with a pet compost pile. Miller brushes by and, as is his wont, sprints straight across the backyard, past the heap, on patrol. Often, squirrels use the stockade fence that runs along the rear of the property as a highline, and he is eternally vigilant in keeping them off his territory.

Perhaps buried deep in Miller's jealous bones is the memory of a time when squirrels were a popular pet. "In the 18th and 19th centuries, squirrels were fixtures in American homes, especially for children," I read on *Atlas Obscura*. While many people captured their pet squirrels from the wild in the 1800s, baby squirrels were also sold in markets. In fact, some suggest the squirrel trade to wealthy urban families helped give rise to the pet-shop industry, today a $70 billion business.

Of course, squirrels being squirrels, in time their allure as domestic pets faded. By the early 1900s they came to be viewed as pests, and now most states have laws that prohibit keeping squirrels and other wild animals at home. Miller must not have gotten the memo.

I take a seat on a patio chair and eye a cicada that's fastened to the top of the front right leg. The thrum of cicadas earlier in the summer was loud enough to drown out the sound of traffic from the nearby interstate. The sight of the bulbous insect, with its iridescent wings and truly bug eyes, reminds me of a boyhood summer in Kentucky, when we captured the lumbering locusts and tied lengths of kite string to their legs to fly them in tight circles above our heads. As cicadas live and die in seventeen-year cycles, that was three generations ago; I hope too long ago for them to hold a grudge.

Actually, this visitor is probably one of the many species of annual cicada who emerge each summer. That likelihood dawned on me the other morning when I stepped out the back door to find a burrow neatly excavated between the flagstone pavers of the patio, hardly a step beyond the stoop. At first I thought the hole was the handiwork of a particularly brazen chipmunk. Similar tunnels pockmark the patio nook behind the barbecue grill; when I wander outside I sometimes see chipmunks ducking into them. Anyway, I was sitting in the chair when an inch-long black-and-yellow striped wasp landed next to the burrow with a cicada in tow. The wasp quickly dragged the gossamer-winged insect down into the burrow, faster than I could switch my cell phone from text to camera.

After some Googling, I found that *Sphecius speciosus*, also called the cicada hawk, is a large, solitary digger wasp that burrows a foot or so deep in sandy soil and provisions its chambered nest with annual cicadas, often captured in flight. The cicada killer paralyzes its prey with a venomous sting so the victim remains alive and fresh until the wasp larvae are ready to feed on it.

The cat is less tolerant of this interloper. While his efforts to bat the buzzing beast out of the sky are half-hearted if not comical, Tuffy is vigilant in defending his turf at ground level, lying in wait under the picnic table to seize any cicada the wasp isn't quick enough to drag into its lair. He'll paw the creature until it's tattered to pieces, which, I suppose, he sees as proof of his primacy over the patio. While not as epic as the battles between predator and prey on, say, the African savannah, the life struggles that play out on the scale of a suburban backyard can be just as gripping and take place, quite literally, at your feet.

"To those who have not yet learned the secret of true happiness, begin now to study the little things in your own backyard." – George Washington Carver

I leave the cicada to its perch—and fate—to plot out a busy weekend ahead. I've set aside Sunday for dispatching my pile, and I have prep work to do beforehand. It's the biggest compost heap I've ever raised, and harvesting it will be a happy but sizable holiday chore.

I read that our region has set a record of sixty straight days of eighty degrees or hotter daily temperatures, with a prediction of

an unprecedented string of ninety-degree days to start the month of September. Welcome to the new not normal. While the sprinkler waters the sections of garden where I prematurely transplanted the ferns and hydrangea, I wade into the pachysandra that surrounds the house like a green moat. If left unchecked, the spreading groundcover would snake its way up under the wood-shingle siding, something I noticed it had begun to do while gathering wayward bits of moss scraped from the roof a few weekends ago.

Pachysandra rips up easily, but it's messy work, and by the time I wend my way around the three sides of my house where the foreign invasive grows to create a bare-ground buffer between the groundcover and the foundation, I'm drenched with sweat. I drag three small tarps' worth over to the brush pile and, in return, sling shovelfuls of raw compost up against the house to cover the newly exposed roots. A flashing of fresh humus against the foundation looks tidy, and the fluffy buffer will make it that much easier to pluck out the racings of next year's growth. Someday I'd like to remove the pachysandra—a.k.a. "the vinyl siding of the garden"—altogether, but what a job that would be. It may have to wait for the day the excavator comes to tear down the house and dig the larger foundation of the McMansion that will surely replace my little cottage.

As soon as the morning dew dries, I haul out the lawn mower and lower the blade a notch. Even though it's only on the middle setting, the mower scalps the parched grass, sending out plumes of dust from under the carriage. The lawn hardly needs mowing, but I want to crop it as low as possible before aerating the turf and top-dressing it with compost.

My poor old lawn; even as I chip away at it, reclaiming parts for plantings, what's left takes a beating each year. The grass is heavily trafficked by me and my garden wanderings, by the dog chasing tennis balls, by kid traffic and deer grazing. The sod I keep on the thirsty side must also compete for water with the superficial roots of the trees that surround and shade it. So every couple of summers I follow the lead of turf-growing pros by aerating the lawn with a rented machine.

The close-cropped mowing done by noon, I make a trip to the equipment rental store. Configured like a walk-behind lawn mower, the core aerator I lug home features a cylinder spiked with hollow steel tubes that rotates when the self-propelled drive is engaged. In front is a large rubber wheel filled with water that, along with a set of detachable lead weights, helps drive the hollow boring tubes into the ground. As they rotate, the tubes extract and extrude soil, each revolution producing a row of fresh plugs about the size of a finger.

Muscling the gas-powered beast back and forth across the yard is a workout, and I have my hands full steering it. But in the time it takes to run a mower across the lawn, the spiked drum has punched thousands upon thousands of four-inch-deep holes into the turf. Littering the ground are that many plugs of soil, each a cross section attesting to the health of the turf and the earth that supports it. Most of the cores comprise a snippet of green grass above a crumbly layer of brown thatch and a tangle of roots still clutching a small cylinder of soil.

Aerating turf helps break up the soil so air and moisture can seep in. The myriad holes will soon be filled with compost, mixed with

the plugs of dirt and grass roots when they break down, along with the grass clippings that have been accumulating over the past few mowings, with much more leaf litter to come. While more superficial than the deep-tine aerating I perform most springs by hand and foot with the pitchfork, this mechanical treatment is far more extensive. I'm sure the trees and garden beds that surround the outside of the yard and flank the house benefit richly, too, as the rain and then snow percolate through this sieve of organic debris to the water table below.

I've rented the machine for four hours, which gives me just enough time to aerate not only my third-acre of turf, but also the Favreaus' yard. With minutes to spare, I trundle the machine over to the Grissoms' too, so Carl can take a pass across his well-worn front lawn. The neighbors help keep my pile supplied with leaves and grass clippings and other organics throughout the year, and it feels good to be able to return a favor.

I rise early on Sunday morning. My day of rest will have to wait until tomorrow's Labor Day holiday, for it's time to dispatch my pile.

Having borrowed snatches of mature compost for various backyard projects over the past few weeks and with next year's batch already gestating in the garbage bin, today's final disbursement comes with more resolve than fanfare. "Care less for your harvest than for how it is shared and your life will have meaning and your heart will have peace," says Kent Nerburn, a writer of Native American narratives.

The simple act of tossing chunks of compost with the hay pitchfork into the wheelbarrow breaks most of the clumps apart, though every so often I stop to pluck away a not-quite-cooked sheaf of

compressed leaves. These I toss into the back corner, building a mini pile that I'll keep in reserve to seed next season's batch. Some will go to top off the garbage can of compost in waiting. Each wad of old leaves is a veritable Dagwood of bacteria, fungi, and microbes.

I alternate hauling loads between the fresh-plugged yard and the new pollinator garden. One for the grass, one for the flowers. It takes a dozen or so scoops with the wide-tined pitchfork to fill the wheelbarrow and an equal number of flings with the spade to disperse the compost in arcing swaths across the yard. It's not heavy lifting, as each load probably weighs forty or fifty pounds, but it is repetitive, and I fall into an easy rhythm. The finest parts of the compost go to the lawn. After every few loads, I stop to rake in the thickest deposits. The tines of the rake further tease the clumps and turf plugs apart, and most of the compost disappears through the blades of grass to fill the holes with fresh humus. The coarser stuff gets upended by the barrow into the new garden; if the rake's not handy, I just kick the compost across the dried-out clods of upturned sod. I'll further mix in this rich, moist soil amendment when I dig holes for the incoming bulbs, plants, and shrubs.

I can cover a couple hundred square feet—a roomful—of lawn with each wheelbarrow load, and after a couple hours, most of the turf is swathed in a scruffy patchwork of tossed raw compost. A neighbor walks by with his young son on a tricycle, looks across the rough mess littering the yard, and asks, "What happened to your lawn?" I explain that it's just temporary and then take a few more minutes to rake in some of the heaviest patches. If screened and sifted humus is the smooth variety of peanut butter, then my compost is very much the chunky style.

Before long, I've covered the new pollinator garden as well, taking fifteen or so loads in all to smother the upended grass in a matrix of dense humus inches deep, the color of the richest dark chocolate. Here's where my pile is finally in its element, arrayed across the bare ground. With its fresh, sharp edgings and uniform, roughly raked covering, this new island continent of compost has the clean, serene look of a Japanese Zen garden. Seems almost a shame to intrude upon it with plantings, but that is just what it's made for.

Back at the heap, I excavate deep into its dwindling reaches, unearthing dank, cold-pressed humus—the really good stuff. I begin making trips with the filled-to-the-brim wheelbarrow to my three closest neighbors. Across the street, Carl has recently renovated the front walkway to his home, creating a garden area lined with granite pavers for his girls to plant. They have yet to backfill with topsoil, so I trundle over several loads of compost to spread across the hard-packed base earth and save on what they'll need to buy from the nursery.

Next door, Danute tends her victory garden, and I dump four loads of compost into a pile nearby for her to mix into the new and already weedy beds. For the Favreaus, I spoon out a fine barrowful of humus as top dressing to prettify their pocket display of annuals.

By late afternoon I figure I've spread enough compost around the yard and neighborhood to fill a full-size pickup bed to the gunwales. A ton or so of crumbly humus is but a small deposit when cast across a patch of turf or dumped straight onto dirt. All that remains of my pile is a rough mixture of rejected forkfuls from those many wheelbarrow loads, plus a few seashells and a stray corncob or

two. Over the next couple weeks, as I do more fall transplanting and find other garden chores to tackle, I'll sift through my mini pile further until it shrinks to a few shovelfuls.

Carl joins me for a beer in the backyard and to make sure I'm coming over for his family barbecue tomorrow. He looks across the lawn to the empty space where my pile has resided for the past ten months and asks, "Where did it go?"

Empty Nest

I arrive home from work midweek and change into my garden uniform—cargo shorts and ratty old t-shirt. Storm clouds gather as I head outside for my usual "compost hour" of unwinding and reconnecting with my small patch of outside world after a day spent inside at a desk in front of a computer screen. "People come to gardening for the refuge of a personal Eden, endlessly complex in its makeup, gloriously simple in its demands," Thomas C. Cooper writes in *The Roots of My Obsession*. As I unplug in my backyard Eden, the forbidden fruit is my Apple iPhone.

Cole called last night to fill me in on his first few days at school. He told me that one of the classes he had signed up for was on sustainability and "ESG." I asked him what that meant. Green energy?

"Geo-engineering—seeding clouds, solar radiation management, desalination. We're going to need fresh water and air—cool air. We're going to have to build new mega cities for climate refugees—like, for 40, 50 million people..." I'd never heard such grown-up talk from him. (And, yes, I had to look up the wonky acronym: Environmental, Social, Governance.) I thought back to our goodbyes outside his dorm, when Cole whispered in my ear, "Thanks, Pop. You helped make me who I am." With that last hug I realized he'd grown so tall that for the first time my shoulders fit under his armpits.

We humans could have reversed course, easily changed paths. Now I fear the compost heap I keep won't amount to a hill of beans compared with the solutions that my son's generation will have to come up with to save the planet and its people. No doubt Cole and his fellow Gen Zers will make their own mistakes in tackling the crises we've left them, but I am sure they will do their best. They will have to.

It's still a surprise to see my pile reduced to a remnant of its former self, an empty nest. Since I wrested several logs out of place to gain access to the heap for its final turn and disbursement, the left side is still akimbo. Now I need to reset them so they can contain the coming deluge of leaves and seaweed and grass clippings. One log, the skinniest of the lot, is rotted and too long to repurpose as a bug hotel in one of the garden beds. I set it on the dolly and trundle it over to my ever-growing refuse pile of tree branches and pulled groundcover. I reset the remaining logs to close ranks, using small, flat rocks to chink them in place so I can safely walk across their tops. As I'm letting more leaves spend the winter atop the perennial beds, I can make do with a slightly truncated heap. And I'll keep the brush pile, loaded with the seeds and pithy stems of the summer's flowers, as an on-site larder and safe harbor for overwintering bugs and critters till spring.

There are other backyard tasks to tackle. I use a hand-crank spreader to cast a small bag of seed across the thinnest patches of my lawn, well on its way to recovering from the core aeration. It flings the tiny seeds outward like sprinkles of rain. I grab a few handfuls to toss directly on the barest spots. Aside from tending my small plot of vegetables and herbs, this is the closest I get to feeling

like a farmer. It may just be in man's nature to cast seed upon the ground. I marvel at how many new living things I'm introducing to my yard with these throws; it must be many thousands.

This is also the time of year to tour the pollinator garden with a small brown paper bag to fill with wildflower seeds. First, the silky milkweed pods, for future monarchs. Then a bag of cosmos, for the bumblebees. "Once you really see one plant's seed, you begin to see seed everywhere," writes Jennifer Jewell in *What We Sow: On the Personal, Ecological, and Cultural Significance of Seeds*. "To know and care for seeds ourselves is one of the most proactive steps we can take to rebuilding our human food systems, our social systems, and the global ecosystems of biodiversity on which we all depend."

I enjoy the cleome most of all, for their profuse flowers and for the small, banana-shaped pods that produce an amazing number of little black poppy-like seeds, to which the doves that flock to my backyard seem addicted. I cull a handful of seeds from the ever-flowering cleome stalks most every evening and scatter them on the patio flagstones for the doves that come at dawn. I can hardly go outside the back door without a dove or three taking flight. Mourning doves are very flappable, and their wings beat noisily as they burst into the air to perch on the tool shed roof, where they bide their time while Miller and I (and the cat) are out and about, before gliding back to the ground to resume their pecking.

Waiting for the rain to arrive, I wander across the yard to rake in the seed, stopping to pick out the inevitable sticks, stones, and shells that expose themselves as the compost bakes in the sun and crumbles into the turf. The largest shells I toss into the vegetable

patch to bolster next year's tomatoes; I pitch the wood chips into the perennial garden beds. The infernal little fruit stickers I wrinkle up between my fingers and flick aside.

I have more backyard chores to attend to over the next few evenings. I've been meaning to improve upon a set of stepping stones I'd laid down several years ago in the grass on the outside perimeter of the vegetable garden, past the patio, where the cleome grows thickest. They've since sunk far below the grass. Whether that's from the weight of footfalls or because the sod has risen, I don't know. But it's time to extend the patio along the flower bed and flank it with a new strip garden of more cleome, for the birds, and more milkweed, for the butterflies. I've already christened this the new "dove walk" for all the cleome seeds that will fall upon it for the birds to pluck at rather than having to find them pecking through the grass. I'll reuse the pavers to edge the new border garden; Carl is gifting me some recycled flagstone he's pulled from his own walkway to make the patio garden for his daughters.

After lifting the pavers, I scrape away the turf along the flower bed, about ten feet long by three feet wide, and toss the scraps, grass side down, to cover the bare floor of my nascent pile, still scruffed from chopping out the infiltrating maple roots. The grassy mat, rich with worms and organic life, will make a fine base layer for the imminent crush of leaves.

To build up the base for the flagstone, I add several buckets of the small stones plucked from the yard this spring and lately from the new pollinator garden. I'm happy to find a purpose for them. For more supplies I head to the town's smallest and rockiest beach.

Besides its contributions of sand and seaweed for my pile and backyard, the local strand also churns up a steady supply of beach brick.

Tumbled by surf into streamlined pieces of all shapes and sizes, the washed-up artifacts of fired clay make fine filler for the walkway within the vegetable garden. I like how their ocher hues contrast with the black dirt. The bricks soak up water like a sponge yet drain like gravel. In spring they absorb the warming sun; on frosty mornings the bricks are rimmed with a coating of ice from moisture sweated out overnight. My favorite find is a pale yellow-orange variety flecked with bits of shell and straw. Surely these pieces date from the pre-industrial era, when brick was formed of hand-dug clay and any old beach sand, then padded with salt marsh grass. They will last forever, more or less. Plus, they're free for the taking—since this maritime bric-a-brac is, technically, discarded manmade refuse from when we valued our beaches differently, I have few qualms about exercising salvage rights over them. And I figure nobody's going to miss the couple buckets of sand I need to finish off the dove walk. The cat in particular loves it when I brush the sand across the flagstones to fill in the cracks, arching a stretched back on the warm, gritty stones. At least until the next rain washes the scratch pad clean.

I also bring home pocketfuls of beach stones, each plucked in stride after catching my eye. I favor polished quartz and garnet and jasper and especially prize the rounded nuggets of chert with their butterscotch luster. Nondescript when dry, these rocks look most fetching when wet, and I toss them under the drain spouts and along the gravel and flagstone walk-up from the driveway to the

side door porch. I like to stand at the edge of the porch overhang when it rains, gazing down at these souvenirs from a hundred walks along the beach. Some are so distinctive I can still remember the day I first picked them up.

Rocks make good eye candy for the gardener, especially in the off-season when green growth and showy colors are absent from the landscape. Any appointments of sculpture, wind chimes, or garden gnomes are a matter of personal taste, of course. Architect Philip Johnson, of the nearby Glass House, calls them "events on the landscape." Aside from the yogurt-flavored setting stone in the corner of the yard and the old tree stump impaled by a pitchfork, my sole *objet de jardin* is a compost heap; such is my eventful life. Although if I could come across a cool bird bath with, say, a splash of lichen on it, one that fits my wallet and the back of the car on the ride home from the estate sale, I'd be a taker.

But I digress. That's the thing beachcombing shares with gardening—filling a bucket with sand or seaweed, a pocket with rounded brick, or a wheelbarrow with fresh compost inevitably invites stray thoughts.

Nothing against gardeners who busy themselves with annual plantings, but I tend to focus on the hardwoods and hardscape of my property, which calls for continual upkeep and the occasional facelift. My goal is to give my yard the best bones I can and to leave it in better shape than I found it, and adding healthy amounts of humus each year is the best way I know how. Another homeowner could, and probably will, change the property as much as I have—the stones I set in place are steps that are fleeting even for me, but the new living soil I've added to the property will remain.

"In American gardening, the successful compost pile seems almost to have supplanted the perfect hybrid tea rose or the gigantic beefsteak tomato as the outward sign of horticultural grace," writes Michael Pollan in *Second Nature*.

Pollan's musings on compost led him to conclude one fall day that "if I wanted to perfect my gardening faith I would have to begin my own compost pile. Which I promptly did... and forgot about it."

"By the time I returned to the compost pile in April, I had read enough about American gardening to know that composting was a pretty silly fetish. It would never produce a beautiful perennial border, just a morally correct one, and wasn't that a little absurd? Well, I guess it is, but when I lifted off the undecayed layer of leaves on top and ran my hand through the crumbly, black, unexpectedly warm and sweet-smelling compost below, I felt like I'd accomplished something great. If fertility has a perfume, this surely was it... this heap of rotting vegetable matter looked more lovely to me than the tallest spike of the bluest delphinium. Right then I realized that, like it or not, I was an American gardener, likely to cultivate in the garden more virtue than beauty." – Michael Pollan

My backyard, and at the heart of it, my compost pile, is a hobby farm of very modest proportions. I've designed and kept it up

according to my wants and whims, to be a playground for my son and me, a robust habitat for native flora and fauna, a family footprint of stored carbon and other essentials that is sustainable and livable. The backyard I tend is a long way from achieving any sort of regenerative, permaculture status, but what its sundry parts and my pile have contributed to it over the past few years, especially the tonnage of new compost, will pay dividends for generations.

Starting Over

On the last day of summer, a bright, sunny Sunday, I will prepare my pile and backyard for the fall.

This season was once known simply as "harvest," from an Old English word for picking or plucking, to reflect an age when farmers gathered their crops for the winter, roughly between August and November. By the late 1800s, "fall" became more common in the U.S. as more people moved into cities, a reference to "fall of the leaf," dictionary.com informs me. Astronomically, of course, the season lasts from near the end of September until mid-December, between the autumnal equinox and winter solstice. Evidently, nowadays it's mostly the Brits who prefer "autumn," from the ancient Etruscan root *autu-*, which refers to the passing of the year.

The hot, dry end to summer has not been kind to the grass I sowed a fortnight ago, though with stints of watering and a passing shower, the freshly aerated lawn has grown in nicely in all but the sunniest spots. I've lavished more water on the new pollinator garden; clusters of mushrooms are sprouting here and there, which I take as a sign the compost I added is spreading its own fecundity down into the turf.

Nature abhors a vacuum, and I a vacant heap, so I turn my attention to the vegetable garden to begin anew. Bound by two corners of

the house and two sidings of six-by-six-inch wood beams and cedar fence posts, along which I've strung a wire fence, the vegetable garden is ringed by a two-foot-wide border of flowers and bisected on the inside by a cross-shaped walk of beach brick. Truth be told, I spend more time each year tending my pile, but the tomato plants, arugula, basil, cilantro, kale, lettuce, and other plantings high and low that thrive in this twenty-by-twenty plot keep the neighbors and me in salad all summer long.

It's time to harvest the riot of spent flowers and vegetables—the greens I planted in spring have mostly all bolted. The fennel has grown tall and lanky and bends over the fence, two of the tomato plants have withered with rot, and the watermelon vines have sprawled across the beach-brick path.

Cucumbers and beans climb along the low wire. Towering sunflowers, spiky cleome, slender cosmos, and sweeping Joe Pye weed bloom in the narrow beds that border the fence. The flowers grow thick enough to do a fair job in keeping the deer at bay, though this summer they've nibbled the vines of cukes and beans that have spread outside the enclosure. Then again, it could be the fat groundhog that has taken up residence under the tool shed. The old saw is true: Plant three times what you need—a third for yourself, a third for the neighbors, and a third for the critters. The biggest and best surprise comes when I tug at the raggedy stems of the potato plants. I'd mostly forgotten about them after burying Danute's cupboard cast-offs this spring, but when I pull the roots from the ground, I find hanging from them a mass of potatoes, of varying shapes and sizes, looking just like the ones I find in the bin at the grocery store.

Most weedy are the strawberry plants that have taken over half the garden. Though they produce a couple weeks' worth of tasty berries each May, the runners and their daughter plants have overwhelmed my small patch. I rip up most, leaving only a few established plants in the back shady corner that I promise myself I will restrict the strawberries to next spring.

Hating to waste so many viable fruit plants, I tease out a dozen or so of the biggest clumps to give to Carl's daughter, who is the garden's main strawberry picker each spring—at least of those the chipmunks don't steal first. He's promised her that next year they'll have their own strawberry patch in their new walkway garden, and these pickings should do the trick.

I tuck the strawberry plants in a bucket of compost and take it across the street, then circle back to fill a wheelbarrow with clumps of black-eyed Susans and coneflowers, excised from the vegetable patch, to further populate the pollinator garden.

The rest of the garden trimmings I pile together and, using the long, weedy fennel plants as a sled, drag the lot over to the heap to layer over the scraps of upturned sod. To this box spring of rubbish I add the most tattered of the purple coneflowers and Joe Pye weed, which I tidy up on a stroll around the garden beds, scouting for more transplants for the pollinator garden. They grow nearly as tall as sunflowers, and their woody stems will make good, fibrous airways beneath the mattress of leaves to come. I leave the stoutest coneflowers standing for the finches, who perch on the waving stems to pluck the *Echinacea* thistles well into winter.

With the stately Joe Pye now a dominant feature of the pollinator garden, I finally look up the source of its curious name. According to Mohican lore, it's the phonetic translation of *Zhopai*, a Native American word for typhoid fever. Joe Pye is the name early colonial settlers in Massachusetts gave to the medicine man who knew how to use the plant to cure them of the deadly scourge. There's a further claim that after his tribe got pushed west from their ancestral home, the settlers forced the sachem to stay behind to continue attending to them. It's a story too on brand, and too tragic, not to be true.

I head back over to the vegetable garden and grip the wrist-thick stalk of a sunflower that had sprouted at the end of the beach brick walkway. The squirrels have already gotten to the foot-wide bloom, bowed heavy with ripened seeds, and now the path is littered with spent hulls. In years past my son and the girls next door would harvest the seeds for a favorite snack-tivity—we washed the seeds, then soaked them in salty brine and roasted them on a baking sheet. The kids enjoyed the idea of fun food coming from the Jack-and-the-Beanstalk plant in the garden. They also liked being able to spit out the shells on the back patio. Now I just hope the squirrels and finches have left a few whole raw shells to seed next year's garden for the birds and pollinators.

I've let this plant, the garden's biggest, go the longest, as I have a special use for it. Tugging the eight-foot-tall stem this way and that, I wrest it from the beach brick. The root ball is the size of a melon, and even after shaking off the stray pieces of brick and dirt, the root and culm must weigh ten pounds. It amazes me how much a plant can grow in a single season.

I trundle the stout sunflower over to my pile and stand it upright in the middle of the other gleanings. It will be the tent pole that supports the heart and lungs of my pile. Around the base of the stalk I nestle the fennel—its hollow stems and filigreed branches give my pile a hint of licorice. I set some of the stouter stalks of coneflower and Joe Pye weed into the base of tangly spent vines, angling them toward the sunflower at the center, like a teepee. I'll soon bury the airy framework under loads of leaves and other stuffings, but I imagine they'll provide some structure before rotting away.

My walkabout leads me to another nurseryman project. Each year, toothpick-size eastern red cedars sprout in the gravel driveway. I tease their taproots from the riprap and tuck them along the old chain-link fence between my lot and the Rosens' house to the west, on either side of a blue spruce I planted soon after moving in. The tallest cedar is now waist high; in years to come, the stout evergreens and the bigger blue spruce they brace will provide a privacy screen for me and perhaps some shelter for the Rosens from the cold winds that swirl in each winter from the northeast. It's the sort of silly little backyard diversion I think most avid gardeners will recognize: a good-faith effort to make use of a garden material in abundant supply—in my case, spindly little cedar sproutings—in hopes that over the years the effort will pay off as a useful garden fixture—a dense hedgerow of living green.

As its presence among the hardscrabble gravel of the driveway attests, red cedar is a hardy, pioneer species that can thrive in poor dry soil as well as wet swampy land. The most widely distributed conifer in the eastern part of North America, they are a common sight in abandoned farm fields and old pastures, solitary sentinels of a landscape in transition. With their rot-resistant heartwood, red

cedar (actually a juniper) makes the best fenceposts; two fashioned from salvaged logs anchor the garden gate. Early American cabinet makers used cedar for fashioning chests and wardrobes, as its fragrant oil repels insects. My grandparents' farmhouse had a full closet with cedar paneling, which made for a very heady secret hideaway. Squirrels strip the aromatic bark to line their nests to ward off ticks and fleas, which is also why cedar shavings make such nice bedding for dog kennels. Their dense foliage and compact frame give shelter to all kinds of birds, butterflies, and small mammals. The red cedar happens to be one of the trees that benefits from increasing levels of atmospheric CO2, a handy trait for these perilous times.

All those reasons make them a good fit for bracketing the spruce. Juniper berries, pale blue and smelling faintly of gin, are a favored winter food for many birds, including the cedar waxwing, which scatters the seeds along with a dollop of their air-dropped fertilizer. You have to admire a tree that has its own bird named after it.

As we parcel out our crowded world into ever smaller pieces, we have fractured the landscape with property lines and forced nature to the fringes. Maybe that's why I love border gardens and the very idea of hedgerows—especially if those meager strips of wild are all that's left to love. We have to make the most of these edges, even if it means pushing the envelope with a neighbor or local utility company. For the life of me (and the planet), I don't know why we would mow more than a fraction of grass along our 160,000 miles of interstate highway, or why we wouldn't plant much of the ground under the 240,000 miles of high-voltage transmission lines with locally native wildflowers and shrubs. In a divided world for people and nature too, not all margins can be clean.

That's the thing about tending a suburban property. Yes, I own the house and its grounds, or more accurately, share ownership with the mortgage lender. But I am fully aware that I am only a temporary caretaker, at least in the timeframe that nature keeps. "Painting is two-dimensional, architecture and sculpture, three-dimensional. But landscapes are four-dimensional, with time being the fourth dimension," says landscape architect Darrel Morrison.

To what he claims scientists are only now formulating as a Theory of Soil, ecologist George Monbiot adds the same wrinkle of time. "The opportunities in a speck of soil can change dramatically from hour to hour. The more complex a system is across space and time, the greater the diversity it can support." My pile is a very souped-up version of that speck of dirt. And only time will tell how a compost heap responds to the longer stretches of extreme heat, the lengthier periods of drought, and the increasing loss of biodiversity that come with climate change and its ever-more unpredictable impacts.

With the ongoing drought and hot, dry days, the trees cling to their still green leaves. As the nights grow cooler and daylight dwindles, they will get the signal to put on their annual show of color, and then the gathering of leaves will begin in earnest. Renovations complete, the heap is ready to receive its annual bounty of a season's growth. Like a chef planning a harvest menu, I plot out the courses I will soon heap upon it. There's a garbage can brimming with a month's worth of kitchen scraps, more spent vines and stalks from the vegetable garden, and seaweed waiting to be carted home from the nearby shoreline. And then, the leaves must fall. For now I can only watch and wait.

"We abuse land because we regard it as a commodity belonging to us. When we see land as a community to which we belong, we may begin to use it with love and respect." – Aldo Leopold

OCTOBER

Color Palette

This year's stretch of fine fall weather—warm days, scant rain, little wind—has allowed the trees that rise from the yard and beyond to cling to their leaves. I take advantage of a balmy Wednesday afternoon to head home early from the office. The end of daylight saving time is still over two weeks away, and as a newly minted empty nester, I have plenty of time to tend to the yard.

In the past week a blaze of fall color has steadily crept up the trunks of the three maple trees along the side road of my corner yard. The crowns are now bright red, the middle branches a golden yellow, and the lower leaves greenish brown. The leaves at the top lead the way in trickling down to the ground.

The sun is the guiding light of my backyard and what grows where and when within it. This being southern New England at about forty degrees latitude—same as Madrid, Naples, and Beijing—the sun's transit across the sky changes dramatically through the year, tracking nearly straight overhead at midsummer before skimming just above the horizon on a short day in darkest winter.

Poets and painters and baseball fans agree with L. M. Montgomery, author of *Anne of Green Gables,* who wrote, "I'm so glad I live in a world where there are Octobers." And with Eudora Welty's take: "My favorite color is October."

"As long as autumn lasts, I shall not have hands, canvas and colors enough to paint the beautiful things I see." – Vincent van Gogh

The palette of autumn is deepening, with dabs of rust and red and all hues of yellow and gold and orange spreading across summer's green canvas. Most of this fall color is still airborne, the collective leaves clinging to their branches, a still life that is anything but still. I freely admit to viewing my pile as a handy metaphor for all manner of notions and things. Of all the comparisons I've found to describe the qualities of the compost heap I keep, the most apt may have been tucked in the fortune cookie that came with the Chinese takeout I picked up the other night for dinner: "*Your life is like a kaleidoscope.*"

Spot on. My pile is a prism of colors and shapes and qualities that change with every moment and twist of the wrist. I have a good idea of how plants grow and why they are green—in a word, because of a magical elixir called chlorophyll—but off the top of my head couldn't explain why or how their leaves change color. It turns out the experts are still somewhat mystified as well. If anyone knows about trees, it's the U.S. Forest Service, the agency that manages nearly 200 million acres of them:

"Three factors influence autumn leaf color: leaf pigments, length of night, and weather, but not quite in the way we think. The timing of color change and leaf fall are primarily regulated by the calendar, that is, the increasing length of night. None of the other environmental influences—temperature, rainfall, food supply, and so on—are as unvarying as the steadily increasing length of night during autumn. As days grow shorter, and nights grow longer and cooler, biochemical processes in the leaf begin to paint the landscape with Nature's autumn palette."

There are three types of plant-based pigments involved in creating autumn color: chlorophyll, which gives leaves their basic green color; carotenoids, which produce yellow, orange, and brown colors in such things as corn, carrots, and daffodils; and anthocyanins, which give color to cranberries, red apples, strawberries, and the like.

The Forest Service further explains that as night length increases in autumn, chlorophyll production slows down and then stops and eventually all the chlorophyll is destroyed. The carotenoids and anthocyanins that are present in the leaf are then unmasked and show their colors.

The amount of moisture in the soil also affects autumn's splendor. "A warm wet spring, favorable summer weather, and warm sunny fall days with cool nights should produce the most brilliant autumn colors," our forester friends assure us.

"In early autumn, in response to the shortening days and declining intensity of sunlight, leaves begin the processes leading up to their fall. The veins that carry fluids into and out of the leaf gradually

close off as a layer of cells forms at the base of each leaf. These clogged veins trap sugars in the leaf and promote production of anthocyanins. Once this separation layer is complete and the connecting tissues are sealed off, the leaf is ready to fall."

And fall they will. And after these Technicolor packets of carbon and minerals are swept up and gathered, my pile becomes a kaleidoscope of ever-changing texture and color. Each load of leaves changes its complexion; the electric yellow of the poplar is swamped by a crush of wine-dark Japanese maple; some oak offerings are scarlet and green, others more ruby red. Some leaves remain stubbornly green, deciding to rot before they rust. Yet more color pours into my pile with every bucket of motley kitchen slop, bag of shredded white paper, bin of green seaweed, and golden straw.

The cold rains and hard frosts of fall and winter wash away the last flickers of autumn's vainglorious hues. Like a spilled tray of watercolor paints, the distinctly primary rainbow of pigments that begins my pile will soon merge into a mush of brown, as dull and uniform in color as it is distinctly rich with decay.

Ruminations

Busy as I keep in tending my pile, garden, and yard, there are plenty of moments when I just take a seat and ponder. Today, it's a canopy of green turning to red and gold on the horizon that frames the swath of sky above. I view this stretch of open space as my property as well, the air rights that allow me to claim passing clouds, lingering sunsets, circling hawks, and fluttering bats as my own.

My favorite perch is a bench set along the front of the tool shed. Made of a slab of burl wood that Michel repatriated from the dump, it was once perhaps meant to be a coffee table and stood against the side of his house for several years before I permanently borrowed it, setting it on two white birch logs too pretty to cleave into firewood. A few paces to the side of this seat is my pile; directly in front is the vegetable garden beside the patio that leads to the kitchen. All in all, a nice shady spot from which to consider my backyard domain.

Miller implores me to play catch, nosing me with a tennis ball dropped in my lap. Aside from the panoramic view, the bench puts me in position to toss balls in two directions across the narrow lawn that flanks the side and rear of the house, both of a distance that's just right for my throwing arm and for the chasing dog to snag one-hoppers.

In the gloaming of a deep summer evening, I can sit and watch the flashing legions of fireflies that rise through the twilight, a light show rising into the ether. Sometimes the chase of a hummingbird flitting among the flower heads keeps me still. Or I watch in awe as a red-tailed hawk swoops low over the roof and passes just overhead, clutching a just-caught squirrel in its talons. Other times I feel the sudden downdraft of a chill, ill wind on a muggy summer day, see the top branches of the trees begin to sway in the gusty breeze, and know I have only minutes to head inside before the fat drops of a dousing thunderstorm plop down upon the ground and bang off roof and driveway.

I see and use my yard, like most gardeners do, as an outdoor living space, a series of interconnected rooms decorated with plants and hardscapes. Although I have a number of nice chairs on the back porch and patio, a comfy, cushioned wrought-iron lounge chair, and even a good sturdy picnic bench in the shade of the backyard, I most often take a perch on a stump or stone. I don't know why, but I suspect these places make me feel more connected to the patch of land I keep in the most organic, down-to-earth way I know. Generally having a pair of clippers or dandelion digger tucked in the back pocket of my grubby jeans or cargo shorts may have something to do with it as well.

The pair of squat logs that begin the walls on either side of my pile also serve as good spots to ruminate. Lately, a pileated woodpecker has been gouging away at the starboard log, grown spongy with rot, searching for grubs. The size of a crow, with a bright crown of red feathers, she is the backyard heap's most exotic visitor and a thrill to see; I always give her space when she comes to inspect my pile.

The rocks that anchor the borders of the flower and fern gardens in three corners of my yard are frequent rest stops too. Each is about as high as a footstool; I know how sturdy these small boulders are because the kids and I dug them out of the ground and rolled them into place. Depending on the day and task, each offers a slightly different perspective on the pleasure of marking time in nature. As the seasons go by and both the garden and I mature, these pauses grow longer and more frequent.

Ruminating is what my pile is all about. From the Latin *ruminat,* meaning "chewed over" or "to chew repeatedly for an extended period"—as in what cows do to cud—the word has long meant "to turn over in the mind" or "to reflect on over and over again, casually or slowly," say the folks at dictionary.com.

Today I'm reflecting not just on the falling leaves and their fate in my pile but also on an intriguing new addition. Beside the heap sits a bucket full of pellet-size poop from a small herd of llamas and alpacas that reside at a nearby living history museum. I'd stopped by to find out more about the organic vegetable garden they keep and to inquire about a Master Composter class they offer. One thing led to another, and that thing is the heavy bucket of llama doo-doo that now sits before me. I find out more on a website kept by Blue Rock Station, a farm in southeastern Ohio:

"Llama manure is... simply put... terrific stuff for your plants. The llama 'beans' as they are often called (as they resemble coffee beans, or rabbit poo, or whatever other 'bean like' thing you care to imagine) break down slowly, releasing their nutrients into your plants. Other advantages include:

- almost no smell (ideal for indoor plants)
- extremely rich in Nitrogen, Phosphorus, and Potassium
- will not 'burn' your plants"

The smooth green nuggets remind me of vitamin capsules, only they are dry rather than filled with oily gel. In fact, a member of a compost forum, permies.com, suggests soaking the llama beans to spur their decomposition, so I am happy to let the bucket sit for a week or so while I wait for the leaves to fall. Showers are on the way, and I'll keep the bucket uncovered to let the coming rain soak in. The llama manure doesn't smell much more than the pleasantly musty sniff of a horse stall. Oh, and in case you're wondering, llamas aren't ruminants. They're camelids, having one less stomach compartment than a cow or sheep, though two more than a horse. I figure the end product is pretty much the same.

I thought I'd be busy by this time of the season with raking. Looking through my compost journal, I know in years past I've gathered heaps of fallen leaves by now, even spread wood chips across the culled flower beds. This fall, the cud for my pile is still clinging to the trees. The cyclical nature of my yard and its workings are increasingly out of whack. So for now I wait and watch and ruminate.

Sometimes I fear this project is less an exercise in sustainability and regeneration than an object lesson in transference, if not outright deflection. Instead of a deep dive into a compost heap, I should be digging into my own psyche. Do I focus on my pile so I don't have to sort out my own rubbish? It took me a long time to heal from the divorce, to move on, to find grace. Having achieved a 100 percent lifetime failure rate in relationships and sent my only child off to college, I've retired to my backyard. It's certainly a refuge, in every sense of the word, a safe harbor to which I've fled. There are certainly worse places to be stuck.

My pile is a friend with benefits. It's made me a better neighbor, a better person, perhaps a better father. It keeps me grounded. In a very basic way, my pile has helped make me who I am. Gardening is my therapy, the backyard compost heap a rustic, restorative retreat from the preoccupations of work and the whatnot of modern life.

"If I were to name the three most precious resources of life, I should say books, friends, and nature. And the greatest of these, at least the most constant and always at hand, is nature." – John Burroughs

The modest dose of natural refuge and respite that my backyard compost pile affords is, in the words of David Strayer, cognitive psychologist at the University of Utah, "a kind of cleaning of the mental windshield that occurs when we've been immersed in nature long enough." In a *National Geographic* article, "This Is Your Brain on Nature," Florence Williams makes the case that "when we get closer to nature—be it untouched wilderness or a backyard tree—we do our overstressed brains a favor."

The latest neuroscience research supports the long-held feeling that nature inspires and soothes the modern mind. All kinds of studies have shown that spending time in green space is an effective

antidote to the health problems associated with spending so much time indoors, from nearsightedness to obesity to depression.

National Geographic gives me a clearer picture of what happens when my brain is on compost. "Korean researchers used functional MRI to watch brain activity in people viewing different images. When the volunteers were looking at urban scenes, their brains showed more blood flow in the amygdala, which processes fear and anxiety. In contrast, the natural scenes lit up the anterior cingulate and the insula—areas associated with empathy and altruism. Maybe nature makes us nicer as well as calmer."

In a conversation with *New York Times* columnist Margaret Roach about her 2020 book, *The Well-Gardened Mind: The Restorative Power of Nature*, Dr. Sue Stuart-Smith says, "People tend to see gardening as a hobby—an activity—but I think it's primarily a relationship. Many gardeners speak of the importance of feeling part of something larger than themselves. This is where the deeper existential experiences in the garden come from, this feeling of being part of the web of life."

"I only half-joke that my favorite place is the compost heap, that eternal-life dimension," Roach confides to Stuart-Smith, a psychiatrist based in England, who created the Barn Garden in Hertfordshire, with her husband, a landscape designer.

"Absolutely," Stuart-Smith answers. "People often describe losing themselves in the garden. Therapeutically, this is important. When the ego falls away and we are at one with a task, we experience a sense of inner calm.

"While gardening brings us into the present, it also has an intrinsic future orientation," Stuart-Smith adds. "The sense of positive anticipation we can feel in working with the natural growth force brings with it a sense of purpose and motivation. Gardening puts us in touch with the transience of life, but it also allows us to feel the continuity of life."

These findings may help explain a random act of composting kindness I found myself performing after the last rainstorm. I wandered out to the curb to find it awash with earthworms, flushed out of the saturated lawn onto the pavement. It seemed a waste that most of these invertebrates would be crushed by tires or drowned in the puddled streams along the gutter, so I took a butter knife and Tupperware bowl out to the street and scooped up dozens of sodden, stretched-out annelids.

I was happy to be working under the cloak of darkness, as it would be difficult to explain the rescue to all but the most committed of gardeners. But depositing the spaghetti bowl of squirming worms onto my pile brought the pleasing thought that these refugees and their progeny would repay me many times over in the coming months. Such is the modest virtue of my pile and the benefits it returns to me. And I now have earned the eternal gratitude of a herd of stranded earthworms. So I've got that going for me, which is nice.

New Vintage

I take the afternoon off, a Friday in the middle of October, to burn up a half-day of paid leave and enjoy another in a string of fine, warm autumn days. October 1st marked the seasonal re-opening of the town's beaches to dogs, on through the end of March. So back to the sea we go, Miller with his tennis ball hanging out the side window and me with an empty plastic tub and three-pronged hand rake.

The dog likes nothing more than to chase a tennis ball across the soft, wet sand into the shallow planes of saltwater at low tide. I stop at the strandline to fill the bucket with a pungent mix of drying seaweed and broken stalks of seagrass. The sand-flecked scrapings are suffused with seashells, the carapaces and claws of crabs, and stray flight feathers, mostly the gray and white quills of seagulls.

The timing is doubly good: After weeks of dry weather, a coming storm system is set to bring rain to the parched region. Bolstered by a harvest moon, the flood tides have already washed up a deep, ragged etching of seaweed and other bounty.

The longer I live by the shore of Long Island Sound, the more I come to respect its riches. And the more I realize that the Sound

I know is a shadow of its former self—its productivity was once legend. I find in Tom Andersen's *This Fine Piece of Water: An Environmental History of Long Island Sound* that "two hundred years after contact, the European invasion had little impact on the estuary's extraordinary productivity. In the eighteenth century, enough lobsters still washed ashore each night from natural die-offs to fertilize the coastal farms of Connecticut." New Yorkers consumed more oysters from the Sound than fish, including the East River oyster, now extinct, whose eleven-inch shell housed seven pounds of succulent flesh.

Andersen quotes from a journal account by a president of Yale University in the early nineteenth century, on the efforts of Long Islanders to improve their agriculture:

"The inhabitants... have set themselves to collect manure wherever it could be obtained. Not content with what they could make and find on their own farms and shores, they have sent their vessels up the Hudson and loaded them with the residuum of potash manufactories; gleaned the streets of New York; and have imported various kinds of manure from New Haven, New London, and even from Hartford. In addition to all this, they have swept the Sound, and covered their fields with the immense shoals of whitefish with which in the beginning of summers its waters replenished. No manure is so cheap as this where the fish abound; nor is so rich; and few are so lasting.

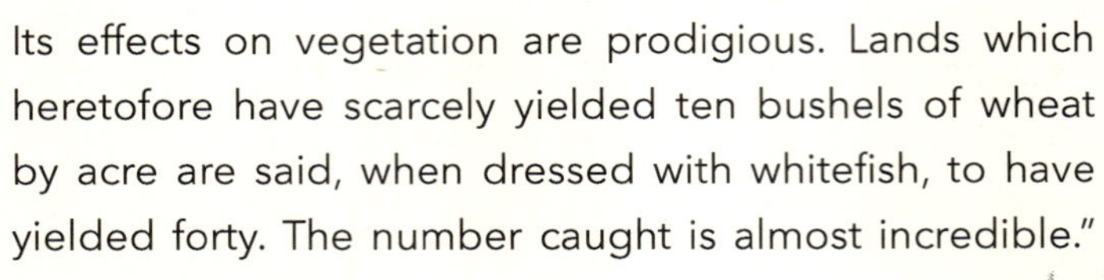

Its effects on vegetation are prodigious. Lands which heretofore have scarcely yielded ten bushels of wheat by acre are said, when dressed with whitefish, to have yielded forty. The number caught is almost incredible."
– Timothy Dwight

Now, having ruinously pulled all these riches from the sea, we disastrously dump our waste into it. As we race headlong into the Anthropocene era, we're producing a planetary load of effluents and manmade spoilings that could and should be easily recyclable. After a toxic algae bloom in the late 1980s, the Sound is struggling to recover even the barest scraps of its once seemingly endless bounty. Though the lobster—once so plentiful that inmates in New England prisons rioted at being served an endless supply—will never return to the ever-warming waters, oyster farms are a growth industry, as is kelp farming; recreational fishing is robust and tightly regulated; beach closings are less common. The summer of 2015 saw the first sightings of whales in the Sound in generations; porpoises and seals are also making appearances, lured from the ocean waters off Cape Cod by growing numbers of baitfish, mostly the oil-rich menhaden.

Menhaden is a corruption of *munnawhatteaug*, which means fertilizer in Algonquian. We know them today as bunker or porgies. Native Americans taught the Pilgrims to plant them with their corn, enabling colonists to coax a crop from rocky New England

soils. Menhaden was also used as a lubricant, replacing whale oil after the Civil War; today most of the catch (the largest by weight in the East Coast fishery) is rendered into heart-healthy omega-3 fish oil, as well as used to produce fertilizers and high-protein feeds for livestock, according to Bruce Franklin, author of *The Most Important Fish in the Sea*.

The bunker were thick this past summer, but the cooling waters this fall led to a die-off, with so many dead fish washing ashore it has kept some local beachgoers at home. Having grown up on those classroom stories of the Indians teaching the Pilgrims to place a fish in each planting hole, I scoop several skeletal fish remains into my bucket. The load weighs about forty pounds, I figure, as it bangs against my hip on the way back to the car. It is ripe enough that I drive home with the windows rolled all the way down, which thrills the dog.

Such a batch of seaweed and organic flotsam is always a prized addition to my pile, especially at this time of year. I will spread it across the base, which began with the layer of spent flower stems and uprooted veggie vines. I've since added the llama beans, more dollops of sod, and a heaping of fresh-cut grass clippings, thick with the rich detritus from when I mowed over the compost dressing. Set on such airy green material, the autumn leaves, I trust, will decompose from below, and the resulting heat and biological activity will filter upward as I continue to layer on fresh additions.

The first tree to give up its leaves is always the red maple that rises beside the road and leans over the Tremblays' front yard, its roots exposed across the patch of hard, compacted dirt where they park

their cars. Danute has already raked the leaves into a tidy pile, a ready-made batch for me to gather onto the old bedsheet.

The fluffy load of crimson and gold easily covers the flattened mound of newly deposited seaweed and mashed-up salt marsh grass. I gaze across the backyard. The lawn, a vibrant green, still grows lushly, and only a shady patch under the big sycamore that lords over the front corner of the yard is flecked with fallen leaves, though not enough to bother yet with raking.

Two scraggly maples on either side of the driveway, their root systems impinged, are always next to drop their leaves, many of which fall on the street. The passing traffic breezes them into long rows along the side of the road, and it takes just a few minutes to rake them up into small collections. Most of these leaves have been pulverized by cars—arboreal roadkill.

Just after I tumble this mix over Danute's maple leaves to finish my composting chores, I'm visited by Carl and his father-in-law, Sal, who comes bearing a most welcome "pile-warming" gift. Carl married into a large family of Sicilian immigrants, and his father-in-law makes wine each fall. This year he bought 250 pounds of red grapes, and today he's brought over a chilled bottle of his first batch. I head inside to grab wine glasses and overhear Sal asking Carl why my grass is still so green.

As I return, Carl is explaining how I spread the compost across the lawn. I set the glasses on a log beside my pile. The cork from the liter bottle comes out with a pop. I pour a tasting for each of us to salute the autumn harvest. The sparkling ruby red wine is fresh and alive.

"It needs to breathe," says Carl.

"It will mature," says Sal, his grandpa-gruff Italian accent nearly as strong as his homemade vino. I taste, we clink glasses, and I look over to my pile and nod in happy agreement.

Tricks and Treats

It's the last day of October, All Hallow's Eve. I read on Wikipedia that "it is widely believed that many Halloween traditions originated from Celtic harvest festivals... it's the time in the liturgical year dedicated to remembering the dead, including saints (hallows), martyrs, and all the faithful departed."

My pile is all about celebrating the faithful departed and hallowed saints, at least those of the vegetative kingdom I lord over. Some Christians historically abstained from meat on the feast of All Hallows' Eve, a practice my pile faithfully honors; others to this day still light candles on the graves of the dead. All in all, it's a most fitting day to devote to the first big cleanup of the yard and in turn to bury the hallowed remains of the season.

I step outside on a pleasant, sunny fall morning to hear the neighborhood abuzz with the sound of small engines, blowing, mulching, and otherwise engaged in the collection of fall leaves. Halloween has become a big holiday. Held at a fine time of year, it's a rare modern tradition that still involves welcoming friends and neighbors to your home, not to mention a large cast of goblins and ghosts and Disney princesses and Marvel heroes. Aside from stringing up Christmas lights or preparing for a Fourth of July barbecue, it's the one time of year when you want your property

looking its best. It's also the last day of daylight saving, and I'm already burning daylight.

Homeowners have different strategies for coping with the seasonal blitzkrieg of leaves. Some spring into action as the first leaf hits the ground, fastidiously sweeping their yards clean of any and all debris and making it a daily habit. Others wait until the final leaf drops before beginning any cleanup. Some homeowners, or their renters, never get around to any seasonal upkeep at all.

I have enough house pride to want my driveway and corner property to be safe for the trick-or-treaters and presentable for the block-partying parents promenading by. I know many by sight from the gatherings near the corner of my front yard each morning, which serves as the neighborhood bus stop, and I have appearances to keep up, after all.

Besides, my pile has been waiting all season for this moment, the day when I raise it from the ground with the first significant contribution from the landscape it nurtures, not just with a batch of freshly fallen leaves, but some special treats as well.

Despite a pelting rain mid-week, most of the leaves of the trees in my yard remain stubbornly aloft. But enough now litter the ground to make a day spent cleaning up a worthwhile yet manageable task. I can't wait to get started.

First, I set out my ad hoc compostables: a barrel of salt marsh grass gathered from the beach, two half-filled bins of scraps from both my kitchen and the neighbors' next door. The big treat is the garbage can filled with scraps from the period when my pile was closed to newcomers.

It's a veritable tinderbox of compost in the making. Though I've topped it to the brim several times since starting it in late August, the bin has settled to about two-thirds full, the white paper frosting on top now stained the color of tea and riddled with red worms. I checked in on my earthworm nursery once to find a mess of worms lining the rim of the lid—probably to escape the brewing anaerobic rot below. There were so many I peeled off a handful and tossed them onto the ground in the shadow of the log walls before stirring the lot with a pitchfork to give the buried worms some breathing room. If the can were mine, I would've drilled a few air holes in it. I imagine bin composters and those who practice vermiculture composting may regularly face a version of this struggle, which makes me glad I keep an open-air heap.

I drag the can around the side log wall and set it in front of my pile. It's too heavy to pick up and dump outright, so I use the pitchfork to spread steaming forkfuls across the growing heap. The pounds and pounds of kitchen scraps, old compost, and rotting paper sink through the freshly deposited leaves, along with many emancipated worms. The dodgy maple seeds, I'm happy to report, are nowhere to be found.

The lawn has grown lush, and I want to cut it not only to add fresh clippings to my pile and chop up this first flush of fallen leaves but also to have the grass short for when the last measure of leaves finally falls. It's much easier to rake leaves off cropped turf.

There was a time when I'd use the spring-tined rake to tease leaves from the fence edges backing the flower beds and from the pachysandra along the base of the house, then fire up the leaf blower to blast any leaves that had fallen in the beds out onto

the grass, where I would mow and mulch them too. But once you learn that such delicate creatures as luna moths and swallowtail butterflies disguise their cocoons and chrysalises as dried leaves, it gets harder to go scorched earth with fall cleanup. "The leaves also serve as a habitat for wildlife including lizards, birds, turtles, frogs, and insects that overwinter in the fallen leaves," advises the U.S. Department of Agriculture. "These living creatures help keep pests down and increase pollination in your garden, so having a habitat for them in the fallen leaves can help to keep them around when you need them the most." So now the perennial beds keep their clutches of wind-blown leaves, and the bugs get their leaf litter to cover themselves through winter.

The mower fires right up and soon I'm trundling along the yard, vacuuming up twenty-two-inch strips of mostly sycamore leaves. The wheels stumble across the tree's many achenes that still dot the ground. Most are too heavy for the whirring blades to slice and dice, but enough are caught to make it seem as though the mower has turned into a popcorn popper, the tough seedballs pinging around the undercarriage. I let the catcher run full, disgorging much of the clippings back onto the lawn before stopping in front of the pile to empty successive loads around the sunflower stalk that rises from the base. After laying down a half-dozen catcherfuls, I dump the rest of the garbage can of starter compost across the top. Adding so much fecund starter to my pile almost feels like cheating.

I finish up the mowing and have just enough daylight left to grab the rake and bedsheet to sweep the side of the street along my driveway clean of leaves, contributing two more loads to my

precocious new pile. Gazing up into the twilight and seeing all the many-hued leaves still aloft, I know there will be much more to come. They'll have a solid foundation on which to rise. Never before have I taken such pains to layer such a diverse supply of raw material right from the start.

At dusk, the yard is a snapshot of peak leaf-peeping season in the suburbs of southwestern Connecticut. Awaiting the first group of tricksters, I find myself wishing to greet a young bee or butterfly, though what a treat it would be to see a superhero version of my pile waddle up to the door. It's not easy anthropomorphizing a mound of humus. Trust me, I know. Not to worry; there is, after all, a whole shelf of children's books about composting, from *Compost Stew: An A to Z Recipe for the Earth* to *What's Sprouting in My Trash?* The kids will be all right. And my young pile has all the makings of a fine heap of hallowed ground.

EPILOGUE

All In

The final page of the calendar for a year in the life of my backyard compost heap has turned. Today is the first Saturday of November, a crisp, frosty morning that promises to warm into a sunny autumn afternoon. Prime time for my pile.

The neighborhood is bustling with activity as I head outdoors to plot the day's pleasant duties. Through a week of seesaw temperatures, a bit of rain, and windy bluster, the trees in my yard have steadily given up the lion's share of their foliage. I look up to see the bare limbs of the maples silhouetted against the brightening blue sky. The lordly sycamore in the corner of the front yard has also dropped most of its leaves, some the size of dinner plates, still green. Cole's young hickory doppelganger in the other corner of the backyard is now surrounded by a footing of amber. Still attached to their petioles, these leaves are too heavy to do anything but drop straight to the ground.

Some leaves in the yard remain aloft—the crimson Japanese maples, privet and forsythia, and more low shrubs and bushes cling to their colorful cloaks. Just beyond my property, the ridgeline of oaks still appears thick and multi-hued, providing a scenic backdrop and the prospect of future wind-driven cleanups.

I reckon I'm halfway done raising my pile off the ground. The first quarter I built up with generous doses of rich greens and heaped upon that base the llama fuel and garbage can of compostables leavened by fluffy layers of salt marsh hay. Last weekend's Halloween cleanup, plus a dosing of kitchen scraps and more, lifted my pile further. Today's sweep will, if the pattern of seasons past holds, make it about three-quarters full. I'll keep after stragglers as long as the weather permits, layering the carbon-loaded latecomers with a steady supply of damp, dark, granular coffee grounds from the local java shop.

Sometimes the fall leaf season ends with a gentle, graceful whimper, like today. Sometimes it ends with a walloping bang. At this moment a decade or so ago, my neighbors and I were coping with the aftermath of Superstorm Sandy, which hurtled up the Atlantic Seaboard on October 29 and slammed ashore just up the Connecticut coast.

Sandy's rain-lashed winds toppled trees across power lines and houses, whipped leaves and branches far and wide, and left the neighborhood without power for the better part of a week. The year before, on the same day, a "freak" snowstorm dumped seven inches of wet, fat snowflakes, burdening the trees still cloaked in leaves with unsustainable weight. The snow melted within the day, but my neighbors and I spent the next week dragging snapped limbs off to the town dump.

In crisis, neighbors pull together, lending portable pumps to drain flooded basements, generators to keep freezers from becoming a total loss, and chainsaws to clear roads and driveways. Picking up leaves was the last thing on our minds. Not today. The whine of

leaf blowers, from backpack commercial crews and homeowners wielding hand-held models, echoes across the neighborhood.

Across the street, Carl is at it as well, and I wave to him as I slowly mulch and mow my own front lawn. He usually rakes all his leaves into a pile before stuffing them into dozens of tall brown bags. Sometimes, I heap a bag or two of his leaves into my wheelbarrow and dump them onto my pile, returning the empty bags for him to refill. But this year he's doing something different.

After blowing the leaves from his thickly shaded yard over the rock wall that fronts the street, he's using his mower to grind up the leaves straight off the asphalt. The result is a windrow of pulverized leaves, minced more finely, and more quickly, than anything I've ever achieved mowing only on the lawn. It hadn't ever occurred to me that the mower would work more efficiently when walked through a pile of leaves splayed out on unforgiving pavement.

I walk over to admire his handiwork, an innovation, Carl explains, born of necessity. He lifts his work shirt to show me the back brace he's wearing and explains that this year there's no way he's doing all the bending and lifting involved in stuffing dozens of paper bags. He makes me a proposition: If I help him drag his roadkill leaves over to my pile, he'll use his mower to help me mulch the rest of my leaves.

It's an offer I can't refuse.

I leave my mower in place in the front yard, grab the large tarp that covers my woodpile, and set it in the street along his property.

Together, we rake and sweep and blow a foot-thick layer of fricasseed leaves, mostly oak, onto the tarp. After gathering up the four corners to drag the tarp across the street, we realize within steps that we've underestimated the heft of all those minced leaves. It's like tugging at a tarpful of sand. We stop to regrip, and Carl worries that his back won't hold up. I take the lead to scooch the fat bag across the backyard, then I clamber up one side of the log wall, dragging the tarp up the slope between us to unfurl it across the top.

The crushed leaves blanket my pile, spewing out from containment like a dry pyroclastic flow of carbon and dust. We collect four more tarpfuls, each load smaller than the previous to accommodate the ever-higher summit. Never before has my pile received such a dense prize of organic matter—the remains of a tree or three, upward of a million leaves, have just exponentially expanded into billions of bits and pieces, which will cook down that much more quickly into new soil.

My pile has become such a spectacle that Carl's older daughter comes out to play, first to toss the tennis ball with the dog, then to scamper up the slippery slope to the summit, where she buries the ball for Miller to nose out. The heap is stiff and stout with crushed leaves, and her small feet hardly sink under the surface. I think back to when Cole was her age, doing somersaults to disappear into the same space. Nowadays, he'd bounce right off.

The take of leaves from Carl's yard and curb far surpasses the haul from my own property. In short order we've spruced up his streetscape and are mowing my lawn in tandem. He mulches in front while I make my way through the layers of sycamore, maple,

and hickory scattered across the back, dispersing a few grass-catcherfuls into the empty corners of the heap.

It's true. It does take a village to make my pile. By midday we've made short work of both our yards and park the mowers. We toast our teamwork with two cold beers from the fridge.

Sipping my beverage on the back porch with Carl, our mowers cooling beside the patio, I consider my pile. It's a stout, round mound, chest high, as it normally is this time of year when fluffed up and airy from mostly whole leaves. I congratulate myself for taking such pains to layer the foundation with as much seaweed and rich greens as I could gather, for now the heap is engorged with a dense load of mulched browns. I couldn't be happier with it.

There's more to come. Carl motions with his beer hand to the Favreaus' across the street, where Pierre has just finished raking a large pile of leaves onto his driveway. Another heap is already set in the corner of his yard. I mention to Carl that Pierre had come by earlier in the morning asking permission to park cars in my driveway tomorrow; he and Joanna are hosting family driving down from Canada on their way to Florida for the winter. As if he needed to ask.

Carl and I both know that Pierre will be hard-pressed to round up and do away with all those leaves on his own. Carl reckons that his back will hold up for a bit more work, so we set down our beers to haul our mowers, tarps, and rakes across the street.

I know now how work crews can be so efficient, for within an hour we've mowed and mulched the Favreaus' yard of leaves and

dispensed with the pile gathered on the driveway, using the big tarp to drag the lot over to my compost heap, which takes it all in.

My pile is now the final resting place of leaves and other castoffs from five homes in our close-knit neighborhood of small lots and helping hands. This backyard heap has become much more than I ever thought it would. And best of all, once again, and in the end, my pile is not finished.

It is only just beginning.

Selected Bibliography

Resources

Below are some suggestions for your compost bookshelf and browser, listed in order of appearance on the preceding pages.

Books

The Gardener's Year, Karel Capek (Modern Library Gardening, 1931)

Let It Rot! The Gardener's Guide to Composting, Stu Campbell (Storey Publishing, 1998)

The Conundrum, David Owen (Penguin Random House, 2012)

How to Make and Use Compost: The Ultimate Guide, Nicky Scott (Green Books, 2010)

The Soil Will Save Us, Kristin Ohlson (Penguin Random House, 2014)

Nature's Best Hope, Douglas W. Tallamy (Timber Press, 2020)

Book of Compost, Mike McGrath (Sterling, 2006)

Green Thoughts, Eleanor Perenyi (Modern Library Gardening, 1983)

Compost, Claire Foster (Cassell, 2005)

The Wood for the Trees, Richard Fortey (Knopf, 2016)

Improving Your Soil: A Practical Guide to Soil Management for the Serious Home Gardener, Keith Reid (Firefly Books, 2014)

Teaming with Microbes: The Organic Gardener's Guide to the Soil Food Web, Jeff Lowenfels and Wayne Lewis (Timber Press, 2006)

Four-Season Harvest, Eliot Coleman (Chelsea Green, 1992)

The Pleasure Garden, Anne Scott-James (Harvard Common Press, 1977)

The Darwin Archipelago, Steve Jones (Yale University Press, 2011)

The Rodale Guide to Composting, Jerry Minnich (Rodale, 1979)

The Art of the Common Place, Wendell Berry (Counterpoint, 2003)

Mycelium Running, Paul Stamets (Ten Speed Press, 2005)

The Nature of Oaks, Douglas W. Tallamy (Timber Press, 2021)

The Lawn: A History of an American Obsession, Virginia Scott Jenkins (Smithsonian Books, 1994)

Sissinghurst: Vita Sackville-West and the Creation of a Garden, Vita Sackville-West and Sarah Raven (Macmillan Publishers, 2014)

Regenesis, George Monbiot (Penguin Books, 2022)

The Roots of My Obsession, Thomas C. Cooper (Timber Press, 2012)

What We Sow: On the Personal, Ecological, and Cultural Significance of Seeds, Jennifer Jewell (Timber Press, 2023)

Around the World in 80 Trees, Jonathan Drori (Laurence King Publishing, 2018)

Second Nature, Michael Pollan (Atlantic Monthly Press, 1991)

The Well-Gardened Mind: The Restorative Power of Nature, Dr. Sue Stuart-Smith (Scribner, 2020)

This Fine Piece of Water: An Environmental History of Long Island Sound, Tom Andersen (Yale University Press, 2004)

The Most Important Fish in the Sea, Bruce Franklin (Shearwater, 2008)

Compost Stew: An A to Z Recipe for the Earth, Mary McKenna Siddals (Tricycle Press, 2010)

What's Sprouting in My Trash?, Esther Porter (Capstone Publisher, 2013)

Websites:

puyallup.wsu.edu/lcs (Linda Chalker-Scott, Washington State University Puyallup)

extension.oregonstate.edu/crop-production/soil (Dan Sullivan, Oregon State University Extension Service)

oldworldgardenfarms.com (Jim and Mary Competti)

margaretrenkl.com

urbanecologycenter.org

plantlife.org.uk

Firefly.org

epa.gov/watersense

scienceline.org

the131school.com (James McSweeney)

compostheaven.com

cals.cornell.edu/school-integrative-plant-science/school-sections/sips-horticulture-section

soiltesting.cahnr.uconn.edu (Dawn)

earthmatter.org

homegrownnationalpark.org

pollinator-pathway.org

joegardener.com

aspetucklandtrust.org

permies.com

awaytogarden.com (Margaret Roach)

Made in the USA
Middletown, DE
15 March 2025

72633007R00203